T0313364

Evolution of Air Interface Towards 5G

Radio Access Technology and
Performance Analysis

RIVER PUBLISHERS SERIES IN COMMUNICATIONS

Series Editors:

ABBAS JAMALIPOUR
The University of Sydney
Australia

MARINA RUGGIERI
University of Rome Tor Vergata
Italy

JUNSHAN ZHANG
Arizona State University
USA

Indexing: All books published in this series are submitted to the Web of Science Book Citation Index (BkCI), to CrossRef and to Google Scholar.

The "River Publishers Series in Communications" is a series of comprehensive academic and professional books which focus on communication and network systems. Topics range from the theory and use of systems involving all terminals, computers, and information processors to wired and wireless networks and network layouts, protocols, architectures, and implementations. Also covered are developments stemming from new market demands in systems, products, and technologies such as personal communications services, multimedia systems, enterprise networks, and optical communications.

The series includes research monographs, edited volumes, handbooks and textbooks, providing professionals, researchers, educators, and advanced students in the field with an invaluable insight into the latest research and developments.

For a list of other books in this series, visit www.riverpublishers.com

Evolution of Air Interface Towards 5G
Radio Access Technology and Performance Analysis

Suvra Sekhar Das

Indian Institute of Technology Kharagpur
India

Ramjee Prasad

Aarhus University
Herning
Denmark

Routledge
Taylor & Francis Group

LONDON AND NEW YORK

Published 2018 by River Publishers
River Publishers
Alsbjergvej 10, 9260 Gistrup, Denmark
www.riverpublishers.com

Distributed exclusively by Routledge
4 Park Square, Milton Park, Abingdon, Oxon OX14 4RN
605 Third Avenue, New York, NY 10017, USA

Evolution of Air Interface Towards 5G Radio Access Technology and Performance Analysis / by Suvra Sekhar Das, Ramjee Prasad.

© 2018 River Publishers. All rights reserved. No part of this publication may be reproduced, stored in a retrieval systems, or transmitted in any form or by any means, mechanical, photocopying, recording or otherwise, without prior written permission of the publishers.

Routledge is an imprint of the Taylor & Francis Group, an informa business

ISBN 978-87-93609-81-5 (print)

While every effort is made to provide dependable information, the publisher, authors, and editors cannot be held responsible for any errors or omissions.

Contents

Preface

कर्मण्येवाधिकारस्ते मा फलेषु कदाचन।
मा कर्मफलहेतुर्भूर्मा ते सङ्गोऽस्त्वकर्मणि॥

(Srimad Bhagwad Gita, Chapter 2, verse 47)

– To work you have the right, but not to the result.
Let not the results of action be your motive, nor let
your attachment be to inaction.

Mankind's quest towards a better quality of life has led to the development of science and technology. Development of science and technology has not only helped improve the quality of life but also it has given rise to an ever increasing demand for improved quality of service which enables our modern living. The different amenities mankind enjoys today are directly or indirectly connected to the rest of the world through some means of communication technology, which brings information to the home/palm of the modern man. In fact availability of communication services are almost a basic necessity for the modernized man.

Such phenomenon has started since the 1990s, when the Internet was made available to the public. With the advent of wireless Internet, mobile communication technology has almost become a part of our existence. This has mainly happened since the rollout of 3rd generation (3G) mobile communication systems and smartphones. The 4th generation (4G) has only made 3G better. Now we are almost knocking at the door of 5th generation (5G) mobile communication networks.

Mobile communication technology has been evolving through various generations. While 3G provided a stream of wireless data, 4G brought in a deluge of infotainment services. This is mainly due to the evolution of radio access technology that have transpired while the mobile communication technology has transgressed through the different generations. While we are entering the 5G era, wireless communication engineers and scientists are preparing solutions to the following new demands, viz. extremely low delay services (tactile Internet with delay in the order of 1ms), provision for dynamically licensed shared spectrum usage, great service in crowd, enhanced mobile broadband (virtual reality being made real), ultra reliable and secure connectivity, ubiquitous QoS, and highly energy efficient networks, devices and services, which have evolved while we were enjoying 3G and 4G. Although the 4G mobile communication systems provide very high speed wireless Internet it does not have the capabilities to meet the above new requirements. Hence is born the new avatar of mobile communication networks namely the 5G.

The above mentioned requirements are expected to be met by several advancement in technologies such as (i) new waveforms or the new radio, (ii) millimeter wave and massive multiple antenna technologies (iii) non orthogonal multiple access (iv) heterogeneous networks such as small cells and device to device communications (v) energy savings in radio access network and (vi) ubiquitous quality of service among others.

In this monograph, we have provided the algorithmic details of the technology solutions mentioned above. In the first chapter, we provide an overview of 5G technology aspects. In the second chapter, we provide several waveforms which were forerunner to the selection of 5G new radio. We also mention the aspects of 5G new radio waveform, its genesis and its variants. In the third chapter, we introduce the non-orthogonal multiple access scheme while in the following chapter we mention the various aspects of mmwave communication technology. The fifth chapter address performance analysis of heterogeneous networks including small cells and device to device communication. The sixth chapter explains energy saving methods for radio access network design through multi-objective optimization. The final chapter introduces analytical performance characterization of QoS for delay sensitive traffic for radio access networks which provide facilities for allocating heterogeneous radio resources to user devices.

The intended audience includes research scholars who are interested to pursue 5G and beyond technologies to bring improvement in performance, which can make these technologies even more commercially viable. It is also expected to provide practicing engineers an opportunity to be familiar with the new technology changes that are expected to be engines of 5G air interface/radio access technology so that they can equip themselves with the new knowledge in a composite manner.

Acknowledgements

The authors are immensely grateful to Dr. Subbarao Boddu, who has put in a great amount of effort in laying the foundation of this document, without which the book would not have seen the light of the day. It was due to his tireless effort that the book could get started and finally get completed.

The authors are highly grateful to all members of IIT Kharagpur, with a special thanks to Prof. Saswat Chakrabarti, the erstwhile head of G.S. Sanyal School of Telecommunications and the director Prof. Partha Pratim Chakrabarti, for providing the necessary support and facility to conduct research work, only because of which the contents in the books could have been produced.

The authors are grateful to Prof. Raja Dutta, head GSSST, for being immensely cooperative in providing all necessary support to ensure the book is completed successfully.

It is extremely important to highlight the contributions of the various researchers who have made the book what it is. The works presented here have emerged mainly from the research work under the guidance of the authors as part of their MS, M.Tech, PhD related work. The chapter-wise contributions of the various researchers involved are given below.

The first author would like to express his gratitude to his parents, Mr. Chandra Sekhar Das, Mrs. Anjali Das, because of whose sacrifices and effort he is able to contribute this little gift to the future researchers of wireless communications. The first author is indebted to his sister, Anusua whose affection has always been with him. He is also greatly thankful to his wife Madhulipa and two children Debosmita and Som Sekhar for bearing through the tough period during the preparation of this work.

Chapter-wise Contribution

2. Waveforms for next generation wireless networks	Shashank Tiwari, Saurav Chatterjee
3. Non-orthogonal multiple access	Priyabrata Parida, Gaurav Nain, Aritra Chatterjee, Debjani Goswami
4. Milimeter-wave communication	Dr. Sandeep Mukherjee, Aritra Chatterjee, Saurav Chatterjee
5. Heterogeneous networks	Dr. Prabhu C, Dr. Asok Dahat
6. Energy-efficient OFDMA radio access networks	Dr. Prabhu C
7. QoS Provisioning in OFDMA networks	Dr. Basabdatta Palit, Dr. Subba Rao Boddy, Dr. Subhendu Batabyal

List of Figures

List of Tables

List of Abbreviations

3G	Third Generation
3GPP	Third Generation Partnership Project
4G	Fourth Generation
AEE	Area Energy Efficiency
APC	Area Power Consumption
ASE	Area Spectral Efficiency
BCCH	Broadcast Control Channel
BLER	Block Error Rate
BE	Best Effort
BS	Base Station
BSs	Base Stations
BWA	Broadband Wireless Access
CCI	Co-Channel Interference
CDF	Cumulative Distribution Function
CDMA	Code Division Multiple Access
DSA	Dynamic Subcarrier Assignment
EDGE	Enhanced Data rates for GSM Evolution
eNB	Evolved Node Base station
ES	Energy Saving
E-UTRAN	Evolved-Universal Terrestrial Radio Access Network
EV-DO	Enhanced Voice-Data Optimized
FFR	Fractional Frequency Reuse
FRF	Frequency Reuse Factor
GA	Genetic Algorithm
GoS	Grade of Service
GPRS	General Packet Radio Service
GSM	Global System for Mobile communication
HeNB	Home eNodeB
HeNB-GW	Home eNodeB Gateway
HeNBs	Home eNodes
HetNet	Heterogeneous Network

IEEE	Institution of Electrical and Electronics Engineers
ICI	Inter-Cell Interference
ICIC	Inter-Cell Interference Coordination
IMT-A	International Mobile Telecommunications Advanced
ITU	International Telecommunications Union
ISD	Inter-Site Distance
KPI	Key Performance Indicator
KRA	Kaufman Roberts Algorithm
LTE	Long Term Evolution
LTE-A	Long Term Evolution-Advanced
MCS	Modulation and Coding Scheme
MIMO	Multiple Input Multiple Output
NLoS	Non-Line of Sight
NSGA	Non-dominated Sorting Genetic Algorithm
OFCDM	Orthogonal Frequency and Code Division Multiplexing
OFDM	Orthogonal Frequency Division Multiplexing
OFDMA	Orthogonal Frequency Division Multiple Access
OPEX	Operational Expenditure
PA	Power Amplifier
PDF	Probability Density Function
PFR	Partial Frequency Reuse
PRB	Physical Resource Block
PSD	Power Spectral Density
PS-RRA	Packet Scheduling-Radio Resource Allocation
QoS	Quality of Service
RAN	Radio Access Network
RF	Radio Frequency
RMa	Rural Macro
RNC	Radio Network Controller
RR	Round Robin
RRM	Radio Resource Management
RT	Real Time
SC-FDMA	Single-Carrier Frequency Division Multiple Access
SE	Spectral Efficiency
SFNs	Single-Frequency Networks
SFR	Soft Frequency Reuse
SINR	Signal-to-Interference plus Noise Ratio
SISO	Single Input Single Output
SLM	Sleep Mode

SNR	Signal-to-Noise Ratio
SON	Self-Organizing Networks
TD-FDMA	Time Division-Frequency Division Multiple Access
TRX	Transmission/Reception
UE	User Equipment
UEs	User Equipments
UMi	Urban Microcell
UTRA	Universal Terrestrial Radio Access
VoIP	Voice over Internet Protocol
WiMAX	Worldwide Interoperability for Microwave Access
WLAN	Wireless Local Area Network
WMAN	Wireless Metropolitan Area Network

List of Important Symbols

$erf\left(\cdot\right)$	Error function, $erf\left(x\right) = \frac{2}{\sqrt{\pi}} \int\limits_{0}^{x} e^{-t^2} dt$
b	Band
I_b	Index of base stations which cause interference in the bth band
c	Center band
e	Edge band
P_{r_i}	Power received from the ith base station in reuse N scheme
$P_{r_i}^{FFR}(r,\theta)$	Power received from the ith base station in FFR
$P_{r_i}^{SFR}(r,\theta)$	Power received from the ith base station in SFR
$\overline{\gamma}_{u,c}(r,\theta)$	Average SINR of the center band user
n_p	Pathloss exponent
$d_{u,i}$	Distance between the user "u" and the ith base station
$\chi_{u,i}$	Shadowing component of user u from the ith base station
$h_{u,i}$	Small-scale fading component
P_n	Noise power
$\mu_{sh-Rayleigh_{pr_0}}$	Mean of channel power from the desired base station in FFR
$\sigma^2_{sh-Rayleigh_{pr_i}}$	Variance of channel power from interfering base stations in FFR
$\mu_{sh-Ray_{pr_0}}^{SFR}$	Mean of channel power from the desired base station in SFR
$\sigma^2_{sh-Ray_{pr_{i_{SFR}}}}$	Variance of channel power from interfering base stations in SFR
P_{T_i}	Transmitted power from the ith base station
$P_{T_c}^{FFR}$	Transmit power for the center band in FFR
$P_{T_e}^{FFR}$	Transmit power for the edge band in FFR
$P_{T_c}^{SFR}$	Transmit power for the center band in SFR
$P_{T_e}^{SFR}$	Transmit power for the edge band in SFR

$I_{c_{FFR}}$	Set of interfering base stations transmitting to the center band in FFR
$I_{e_{FFR}}$	Set of interfering base stations transmitting to the edge band in FFR
$I_{c_{SFR}}$	Set of interfering base stations transmitting to the center band in SFR
$I_{e_{SFR}}$	Set of interfering base stations transmitting to the edge band in SFR
ρ_{p_c}	Power spectral density over the center band
ρ_{p_e}	Power spectral density over the edge band
γ_{th}	SINR threshold
α	Bandwidth partitioning ratio
ρ_p	Power ratio
β	Bandwidth efficiency loss
η	SNR efficiency loss
$\Delta_{f_{sc}}$	Subcarrier bandwidth
f_s	Oversampling factor
$\mathbb{B}$	System bandwidth
γ_k	SINR of the kth class of users
$N_{k,b}$	Number of carriers required by class k users when in band b
$N_{sc}(k)$	Number of subcarriers required for class "k"
n_{sc}^{occ}	Number of subcarriers occupied for class "k"
ρ_k	Traffic intensity for class "k"
$o_p(n_{sc}^{occ})$	Occupancy probability
$P_b(k,b)$	Blocking probability of the kth class of users in band b
P_b	Blocking probability
$\overline{P_b}(b)$	Average blocking probability of band b
$P_b(k)$	Blocking probability of subcarrier class "k"
$P_{b_{th}}$	Blocking probability threshold
s	Value of the received metric (received user spectral efficiency)
$\bar{s}_{ref}$	Reference mean cell capacity
GoS_c	Grade of service for the center band
GoS_e	Grade of service for the edge band
w_{GoS}	Weighted grade of service
w_{GoS}^b	Weighted average grade of service in band b
w_{GoS}^{FFR}	Weighted average grade of service in FFR

w_{GoS}^{SFR}	Weighted average grade of service in SFR
F_{ec}	Metric for GoS fairness
F_u	Measure of proportion of satisfied users
$F_c(\gamma)$	Probability of selecting the center band
$F_e(\gamma)$	Probability of selecting the edge band
K	Number of user classes
K_b	Number of classes in band b
$N_{ch,b}^a$	Number of channels available in band b
N_{sc}	Total number of subcarriers
$N_{sc,c}$	Number of subcarriers in the center band
$N_{sc,e}$	Number of subcarriers in the edge band
F	Noise figure
N_0	Noise power spectral density
$\overline{BW}$	Average bandwidth required by each user throughout the cell
$p(\gamma)$	Probability density function of SINR
$N_{avg,b}^r$	Average number of subcarriers required to make a cell in band b
$b_u^r(r,\theta)$	Effective bandwidth required by a user at a location
R_u^r	Data rate requirement of a user
$\gamma_{u,b}(r,\theta)$	SINR of a user at a location in band b
N_u	Total number of users in a cell
$N_{u,c}$	Number of users in the center band
$N_{u,e}$	Number of users in the edge band
$b_{u_{A_c}}$	Area averaged bandwidth required for a center band user
$b_{u_{A_e}}$	Area averaged bandwidth required for an edge band user
$P_c(r,\theta)$	Probability of a user at a location being in the center band
$P_e(r,\theta)$	Probability of a user at a location being in the edge band
P_{A_c}	Area-averaged probability of selecting band $b=c$
P_{A_e}	Area-averaged probability of selecting band $b=e$
ρ_s	Total offered traffic intensity (Erlang capacity) in a cell
$N_{ch_{avg,c}}^{al}$	Number of channels allotted in the center band
N_{sc}^r	Total number of subcarriers required to make a call
$S.E$	Spectral efficiency threshold point
$\overline{SE}_{FFR}$	Mean cell spectral efficiency of FFR
$\overline{SE}_{SFR}$	Mean cell spectral efficiency of SFR
$\overline{\gamma_c}$	Center band SINR averaged over Rayleigh fading
γ_{max}	Maximum value of SINR

γ_c	Center band SINR
γ_e	Edge band SINR
$\gamma_{max,e}$	Maximum value of edge band SINR
$\gamma_{max,c}$	Maximum value of center band SINR
F_γ	CDF of SINR in a cell
$F_c(\gamma_{i+1})$	CDF of SINR of the $(i+1)^{th}$ class user class of the center band
$F_e(\gamma_{i+1})$	CDF of SINR of the $(i+1)^{th}$ class user class of the edge band
N_i	Number of subcarriers in the ith ring
q	Allowed % reduction in aggregate throughput

1

Introduction to 5G

1.1 Introduction

Wireless communication has changed the social fabric of humanity like no other technology. In a few decades, it has marched steadily into the fourth generation and is preparing to appear in its new avatar the 5G. It is reported in various forums and reports [sta], [thi], [tel] that soon there will be few billion subscribers of 4G worldwide, when it was approximately one million in 2010. Even a developing nation like India is currently world's third largest market for 4G LTE smartphones, over taking Japan, South Korea, and all European countries. Among several notable technological improvements, 4G-LTE allows users to download a feature film in about 10 minutes which took nearly 30 minutes in 3G systems. While the world is reeling under accelerating download speeds from 2G to 4G, researchers around the globe have already started preparing the roadmap for 5G. This monograph is about some of the important technology developments especially in the radio access technology/radio access network that are expected to drive 5G and beyond. This comes at an appropriate time when 5G is getting formulated. The time line of 5G development in 3GPP is shown in Figure 1.1. When consumers are yet to explore 4G to its brim, 5G has already started making noise through several world events such as the IEEE 5G summit [5Gs], about exploiting its potential to realize the following.

- Remote surgery:- Virtual low latency.
    - Doctors can be in remote location from the patient, thus enabling transfer of expertise over great distances. Now patients have to fly to specialists, which is often not feasible on health and economic grounds.
    - One may also imagine driving near one's home from the airport while in office.
- Industrial applications:- Robotic assembly is only overseen by human. Touch and feel may be brought miles away by this.

IMT 2020- Timeline: 3GPP

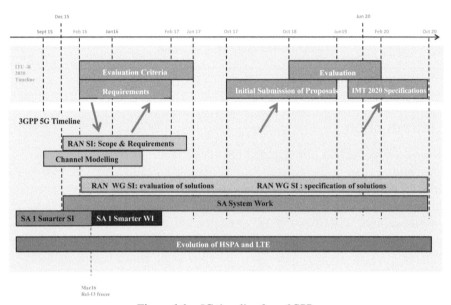

Figure 1.1   5G time line from 3GPP.

- Self-driving cars:- Self-driving cars which require communication and signaling among different vehicles and city infrastructure can be enabled by 5G.
- Drones:- The true potential of drones is expected to be unraveled by the use of 5G technology.
- Virtual reality:- Immersive experience and chat with live streaming virtual world are expected to be made feasible due to 5G.
- Guaranteed service:- Service level agreements (SLAs) for mission critical business and applications are to be supported by 5G.
- Home broadband:- A wire-free home with broadband multimedia services in a smart home is not far away.

To make the above happen, technologically 5G will have to support the following scenarios.

- **Enhanced mobile broadband**: That is, speed like fiber but at a much lower cost per bit. It means to include ultra-high fidelity media anywhere (i.e., fiber like performance without wires). It would enable immersive virtual reality, collaborative interactive education among others. It is also referred to as always connected augmented reality. Physical distances will shrink like anything.

- **Mission critical control**: It is often defined as a new command and control service when failure is not acceptable under normal (non-disastrous) conditions. It has to provide ultra-reliable, low latency links.
- **Massive Internet of things**: The massive Internet of things is usually described as one that needs to provide intelligently connected, virtually anything, anywhere with optimized power and cost.
- **5G NR**: It is the global 3GPP new radio standard. The air interface should be scalable across all services and spectrum (>6 GHz, <6 GHz, <1 GHz). It is expected to be some optimized orthogonal frequency division multiplexing (OFDM)-based waveforms. The air interface is expected to accommodate new spectrum sharing paradigms across licensed spectrum, shared licensed spectrum, and unlicensed spectrum. The air interface is also hoping to include mm wave spectrum bands.
- **Uniform QoS**: More uniform QoS experience (SLA based) is required to be supported along with immersive connected traveling in driverless cars as well as in in safety applications. Several mobility conditions are required to be included which 5G is going to support, for example: high mobility (500 kmph), dense environments, indoors, virtually everywhere.

The scenarios of 5G described above are broadly depicted in Figure 1.2.

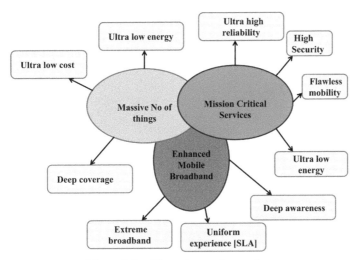

Figure 1.2 5G services and scenarios.

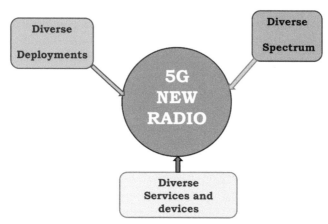

Figure 1.3   Design challenges in 5G.

In order to realize the above, 5G is required to scale down cost and power and also provide faster mobile broadband with uniform QoS. It should deliver new levels of reliability and latency for mission critical services. These extreme variations require a network which is scalable and adaptable. The design challenges for 5G NR are depicted in Figure 1.3.

1.2 Development of LTE toward 5G

While all the above are yet to arrive, LTE is expected to grow beyond its present form and support the following while leading the journey toward 5G.

- Carrier aggregation
- MU-MIMO several layers
- Unlicensed LTE
- Shared license access
- 256QAM
- hotspots
- Device to device (D2D)
- Off-load.

The following are some of the aspects that LTE-Advanced is expected to establish toward its journey in the path of 5G new radio (NR):

- 10 X experienced throughput
- 10 X decreasing E2E latency

- 10 X connection density
- 3 X increasing SE
- 100 X traffic capacity
- 100 X network efficiency.

A representative picture of the above is expressed in Figure 1.4.

Some of the important areas which still require attentions in the development of LTE are briefly described below.

1. Energy efficiency: In developing countries, a good percentage of cell sites are beyond power grid. They require renewable energy. Further, it is interesting to note that more than 80% of the energy spent on operating a network is used for activities other than transmission of bits. It is also observed that a large fraction of energy is consumed by base stations to transmit system broadcast even when there are no data. It is also noted that highly loaded networks transmit information for 20–25% of the time. Interestingly, it has also been reported that more than 70% of the traffic is carried by 30% of BS and the rest of 70% of base stations carry less than 30% of traffic. The reasons for these are that traffic is

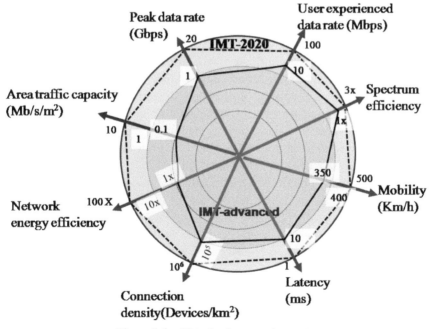

Figure 1.4   5G technology requirements.

not equally distributed between BSs as well as traffic is not equally distributed over a 24-h period. All these suggest need for methods to save energy for the radio access network, which is addressed in this book as well.

2. Evolved HD voice for LTE: It is reported that in 2009, data traffic overtook voice as the dominant traffic in mobile networks, and there is a stable trend of data traffic growth and almost flat voice traffic development. However, it is also reported that voice services still play an important role, since by definition, they are part of mobile and fixed telephony, and people intuitively expect voice service quality to be as high as possible. According to an Ericsson Consumer Lab report, some of the most important drivers of user satisfaction with mobile operators are: network speeds, mobile coverage, and clarity of voice calls. While HD voice is currently being rolled out globally, enhanced voice service is being standardized. The range of human hearing covers frequencies from 20 to 20,000 Hz, while human voice spans a frequency range of at least 50 to 12,000 Hz. However, traditional phone calls (both fixed and mobile) transfer sounds in a smaller spectrum band of roughly 300–3400 Hz. Consequently, the voice quality of today's calls is limited, which requires future networks to address QoS requirements of such calls. The issues related to resource provisioning of such real time traffic, which may coexist with other traffic, are addressed in this book.

3. Cloud RAN: Increasing traffic demand, limited spectrum availability, and mass adoption of mobile broadband are challenging traditional ways of building cellular networks. In this new operating situation, mobile operators are looking for new ways to increase network capacity and coverage while reducing time to reach market with new services while achieving lower expenditure. Cloud RAN architectures are expected to provide support to meet these increasing traffic demands through the use of network functions' virtualization techniques and data center processing capabilities, which allows for resource pooling, scalability, and layer interworking.

1.3 Technologies Drivers for 5G

Some of important technologies that are expected to be in the core of 5G are captured in Figure 1.5, which is briefly explained below.

- New waveforms: New requirements for fifth-generation mobile networks (5G) have triggered the search of the waveforms beyond OFDM.

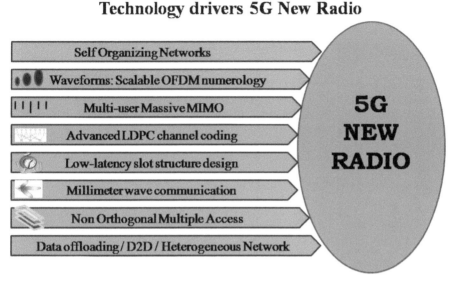

Figure 1.5   Technology drivers for 5G.

In recent years, many waveforms that are primarily different form of filtered multi-carrier transmission techniques have been proposed for 5G systems. Each has its own merits and demerits. In this book, we describe them.

- Radio access network evolution: The radio access network has evolved from macro cell coverage to
    - Heterogeneous network which included small cells,
    - D2D communication, and
    - Non-orthogonal multiple access

  which are addressed in this text.
- QoS measurement and fair service provisioning: The evolution in mobile phones over the recent years has diversified the nature of mobile applications with some requiring high data rates and strict QoS constraints. It is well experienced that present day mobile networks are not able to provide ubiquitous QoS in the coverage area, which is one of the important requirement of next generation systems. In this book, we devote one section which explains methods for QoS service provisioning for ubiquitous QoS experience.
- Mm-wave communications: One of the technologies that have been suggested to meet the requirements of future wireless networks is

millimeter-wave (mmW) spectrum. We dedicate one of the chapters of this book on mmW communication technology. We explain the mmW propagation channel conditions as well as provide an estimate of performance under such channels. We also include new insights such as LoS factor as well as spatial correlation values for performance evaluation of mmW communication systems. Our presentation includes performance of MIMO systems in mmW channels as well.

2

Waveforms for Next-Generation Wireless Networks

2.1 Introduction

Fourth-generation (4G) communication systems [STB11] are designed to provide very high data rate connections to users. The next generation wireless (5G) wireless communication technologies [OBB+14, ABC+14, GRA15] have to serve new application scenarios such as machine type communication (MTC) [DSG+15], cognitive radio (CR) [HWWS14], great service in crowd, tactile Internet [Fet14], etc., which have a wide variety of different requirements, than 4G , to be met. In MTC, the number of devices to be connected is enormous, and hence it requires relaxed synchronization and low-cost device architectures [WKtB+13]. Tactile Internet requires very low latency (1 millisecond) and very high reliability (99.9999%) [Fet14]. CR applications which exploit spectrum in an opportunistic manner need very low out-of-band (OoB) leakage −65 dB [DF14] and flexible operational bandwidth.

Orthogonal frequency division multiplexing with a cyclic prefix (OFDM-CP) is *de facto* waveform for 4G due to its low complexity and capability to combat frequency selective fading channel (FSFC). However, its pulse shape makes OFDM-CP vulnerable to frequency offsets, while also generating significant OoB leakage. These limitations in OFDM-CP have led researchers to find appropriate waveforms for 5G. Based on the above requirements, many existing and new waveforms have been considered for 5G, like filter bank multi-carrier (FBMC) [Bel10, CEO02, Sal67a, ST95], generalized frequency division multiplexing (GFDM) [MGK+12], unified filtered multi-carrier (UFMC) [VWS+13a], block inverse Fourier transform precoded GFDM (BIDFT-GFDM) [TDB15], DFT precoded GFDM [DT15], etc.

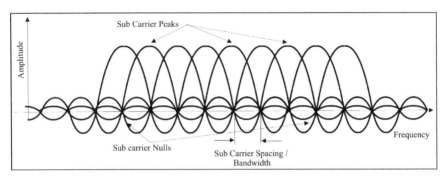

Figure 2.1   Orthogonal subcarriers in OFDM.

Descriptions of the above-mentioned waveforms which play a vital role in paving the path for future radio access technologies are provided in this chapter.

2.2 OFDM

Orthogonal frequency division multiplexing (OFDM) is a multi-carrier modulation technique. OFDM is an advanced form of frequency division multiplexing where the frequencies multiplexed are orthogonal to each other and their spectra overlap with the neighboring carriers. The frequency domain view of the signal is shown in Figure 2.1. The peak of one subcarrier coincides with the nulls of the other subcarriers due to the orthogonality. Thus, there is no interference from other subcarriers at the peak of a desired subcarrier even though the subcarrier spectrum overlaps. It can be understood that OFDM system avoid the loss in bandwidth efficiency prevalent in system using non-orthogonal carrier set. This brings in a huge benefit in spectral efficiency for OFDM systems over earlier systems.

Let us begin by considering a block of QAM modulated symbols $\mathbf{d} = [d_0 \ d_1 \cdots d_{N-1}]^{\mathrm{T}}$. Let us assume that data symbols are independent and identical, i.e., $E[d_l d_l^*] = \sigma_d^2$, $\forall l$ and $E[d_l d_q^*] = 0$ when $l \neq q$. Let the total bandwidth B be divided into N number of subcarriers where the symbol duration $T = \frac{N}{B}$ second and $\frac{B}{N}$ Hz is the subcarrier bandwidth. The transmitted OFDM signal, $x(n)$, $n \in [0, N-1]$, can be given as,

$$x(n) = \sum_{r=0}^{N-1} d_r e^{j2\pi \frac{rn}{N}}. \tag{2.1}$$

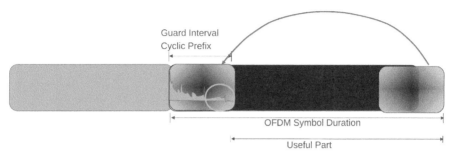

Figure 2.2 Illustration of the role of CP addition. It can be observed that ISI from adjacent symbol is absorbed in CP duration.

Alternatively, the transmitted signal can be given as $\mathbf{x} = \mathbf{W}_N \mathbf{d}$, where $\mathbf{W}_N$ is the N-order inverse discrete Fourier transform (IDFT) matrix. Hence, the transmitter can be implemented using N-point IDFT of complex data symbols, $\mathbf{d}$. IDFT can computed using the low-complexity radix-2 fast Fourier transform (FFT) algorithm. When the OFDM signal is passed through a multipath channel, the symbol spreads in time. To avoid inter-symbol interference (ISI), a cyclic prefix (CP) of length L_{CP} is added. Figure 2.2 illustrates the role of CP. The transmitted signal for OFDM-CP can be given as,

$$\mathbf{x}_{cp} = [x(N + L_{CP} - 1) : x(N - 1); \mathbf{x}]. \tag{2.2}$$

2.2.1 Channel

Let $\mathbf{h} = [h_1, \ h_2, \cdots h_L]^{\mathrm{T}}$ be L length channel impulse response vector, where h_r, for $1 \leq r \leq L$, represents the complex baseband channel coefficient of the rth path [PS08], which is assumed to be zero-mean circular symmetric complex Gaussian (ZMCSC). The channel coefficients related to different paths are taken to be uncorrelated. Also, $N_{cp} \geq L$ is considered. The received vector of length $N_{CP} + NM + L - 1$ is given by,

$$\mathbf{y}_{cp} = \mathbf{h} * \mathbf{x}_{cp} + \boldsymbol{\nu}_{cp}, \tag{2.3}$$

where $\boldsymbol{\nu}_{cp}$ is the AWGN vector of length $MN + N_{cp} + L - 1$ with elemental variance σ_{ν}^2.

2.2.2 Receiver

The first N_{cp} samples and last $L - 1$ samples of $\mathbf{y}_{cp}$ are removed at the receiver, i.e., $\mathbf{y} = [\mathbf{y}_{cp}(N_{cp} + 1 : N_{cp} + MN)]$. Use of cyclic prefix converts linear channel convolution to circular channel convolution when $N_{cp} \geq L$ [STB11]. The MN length received vector after the removal of CP can be written as,

$$\mathbf{y} = \mathbf{H}\mathbf{W}_N\mathbf{d} + \boldsymbol{\nu}, \tag{2.4}$$

where $\mathbf{H}$ is the circulant convolution matrix of size $MN \times MN$, which can be written as,

$$\mathbf{H} = \begin{bmatrix} h_1 & 0 & \cdots & 0 & h_L & \cdots & h_2 \\ h_2 & h_1 & \cdots & 0 & 0 & \cdots & h_3 \\ \vdots & & \ddots & & & & \\ h_L & h_{L-1} & \cdots & \cdots & \cdots & \cdots & 0 \\ 0 & h_L & \cdots & \cdots & \cdots & \cdots & 0 \\ \vdots & & \ddots & & & & \\ 0 & 0 & & h_L & \cdots & \cdots & h_1 \end{bmatrix}, \tag{2.5}$$

and $\boldsymbol{\nu}$ is the WGN vector of length MN with elemental variance σ_ν^2.

As $\mathbf{H}$ is a circulant matrix, $\mathbf{H}$ can be factorized as $\mathbf{H} = \mathbf{W}\boldsymbol{\Lambda}\mathbf{W}^{\mathrm{H}}$. $\boldsymbol{\Lambda} = \sqrt{N} \times diag\{\mathbf{W}_N^{\mathrm{H}}\mathbf{h}\}$, where, $\mathbf{h}$ is the first column of $\mathbf{H}$. This implies that $\boldsymbol{\Lambda}$ holds channel frequency samples which are sampled at the rate of $\frac{B}{N}$ samples per Hz. Hence, received vector after the removal of CP can be written as,

$$\begin{aligned} \mathbf{y} &= \mathbf{W}_N\boldsymbol{\Lambda}\mathbf{W}_N^{\mathrm{H}}\mathbf{W}_N\mathbf{d} + \boldsymbol{\nu} \\ &= \mathbf{W}_N\boldsymbol{\Lambda}\mathbf{d} + \boldsymbol{\nu} \end{aligned} \tag{2.6}$$

Further, N-point FFT of $\mathbf{y}$ is computed, i.e., $\mathbf{z} = \mathbf{W}_N^{\mathrm{H}}\mathbf{y} = \boldsymbol{\Lambda}\mathbf{d} + \mathbf{W}_N^{\mathrm{H}}\boldsymbol{\nu}$. The time–frequency diagram of the OFDM signal in Figure 2.3 shows the difference between single and multi-carrier systems with respect to the symbol duration when compared against the channel impulse response. Single carrier systems have a symbol duration which is decided by the sampling period of the system. When the channel impulse response is larger than this period, there is ISI. The whole bandwidth is split into a set of parallel orthogonal substreams each of which has a long symbol duration. The symbol duration becomes significantly greater than the channel impulse response length. This makes each stream, i.e., each subcarrier, experience only flat fading. Hence, a simple one tap equalizer per subcarrier can be used to equalize the effect of

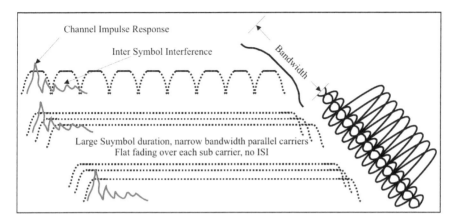

Figure 2.3 Time–frequency representation of an OFDM signal.

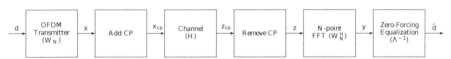

Figure 2.4 OFDM transceiver.

the channel. The effect of the channel can be equalized using a zero forcing (ZF) equalizer. The equalized OFDM vector $\hat{\mathbf{d}}$ can be written as,

$$\hat{\mathbf{d}} = \boldsymbol{\Lambda}^{-1}\mathbf{z} = \mathbf{d} + \boldsymbol{\Lambda}^{-1}\mathbf{W}_N^H\boldsymbol{\nu}. \tag{2.7}$$

The OFDM transceiver can be understood in the light of Figure 2.4.

2.3 5G Numerology

OFDM with different numerology has been accepted as 5G waveforms by 3GPP [3GP18]. 5G waveform is refereed as new radio. Numerology refers to subcarrier bandwidth, CP length, and guard band. 5G supports two frequency ranges, namely, (1) FR1 (450–6000 MHz) in microwave frequency range and (2) FR2 (24250–52600 MHz) in millimetre frequency range.

Definition of channel and transmission bandwidth is given in Figure 2.5. Different user equipment (UE) channel bandwidths as well as subcarrier bandwidths are supported within the same spectrum for transmitting to and receiving from UEs. Channel bandwidths of 5, 10, 15, 20, 25, 30, 40, 50, 60,

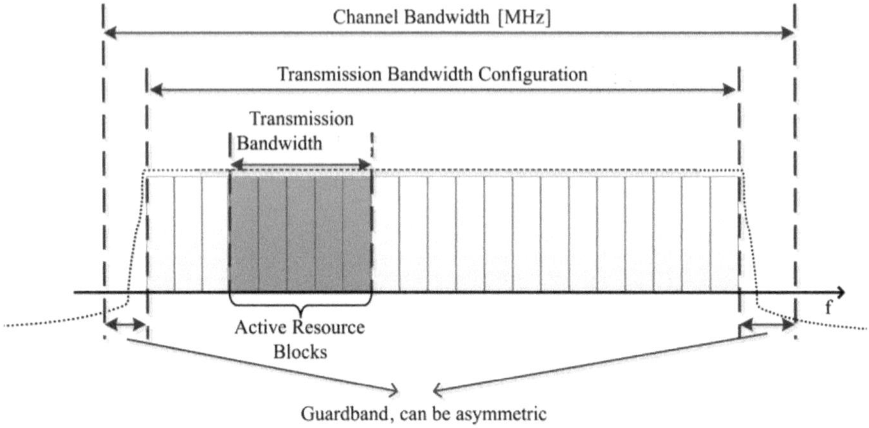

Figure 2.5   Definition of channel and transmission bandwidth.

80, and 100 MHz are supported in FR1. Maximum transmission bandwidths allowed for FR1 and FR2 are given in Tables 2.1 and 2.2, respectively.

2.3.1 Genesis

Use of different subcarrier bandwidths for OFDM was first proposed in 2005 [DR05, DCP07, DCP08]. Figure 2.6 illustrates the concept of variable sub-carrier bandwidth (VSB) in an OFDM system. VSB OFDM implementation can be based on band division multiplexing as in Figure 2.7. The entire available bandwidth may be divided into sub-bands with different subcarrier bandwidths in each sub-band, for example, a 100 MHz may be divided into chunks of 20, 10, or 5 MHz. Each sub-band can be operated on by an IFFT with a different number of subcarriers. The UE is assumed to require only one type of subcarrier bandwidth and hence will operate on only one subband. Therefore, only one programmable FFT is needed for the UE. With a changing requirement of the subcarrier bandwidth, the clock and the FFT size of the programmable FFT may be dynamically configured. At the BS, as many FFTs may be used as there are different types of subcarrier bandwidths.

2.3.2 Implementation

In this section, a different system architecture for implementing adaptive subcarrier bandwidth (ASB) will be described. This kind of architecture can be very well suited for OFDMA systems. For analysis, we shall consider

Table 2.1 Maximum transmission bandwidth configuration for FR1

Channel bandwidth (MHz) →	5	10	15	20	25	30	40	50	60	80	100
Subcarrier bandwidth ↓ (KHz)	N_{RB}	N_{RB}	N_{RB}	N_{RB}	N_{RB}	N_{RB}	N_{RB}	N_{RB}	N_{RB}	N_{RB}	N_{RB}
15	25	52	79	106	133	TBD	216	270	NA	NA	NA
30	11	24	38	51	65	TBD	106	133	162	217	273
60	NA	11	18	24	31	TBD	51	65	79	107	135

TBD means to be decided and NA means not applicable.

Table 2.2 Maximum transmission bandwidth configuration for FR2

Channel bandwidth (MHz) →	50	100	200	400
Subcarrier bandwidth ↓ (KHz)	N_{RB}	N_{RB}	N_{RB}	N_{RB}
60	66	132	264	NA
120	32	66	132	264

NA means not applicable.

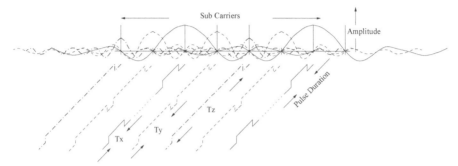

Figure 2.6 Time–frequency description of VSB OFDM.

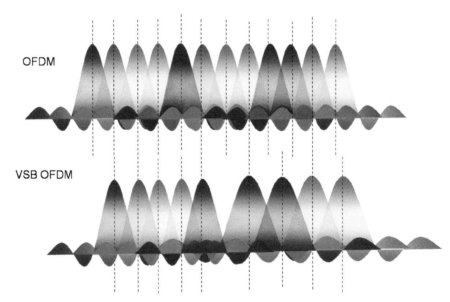

Figure 2.7 Illustration of VSB OFDM. As can be seen, VSB OFDM accommodates different subcarrier bandwidths whereas the subcarrier bandwidth is fixed in OFDM.

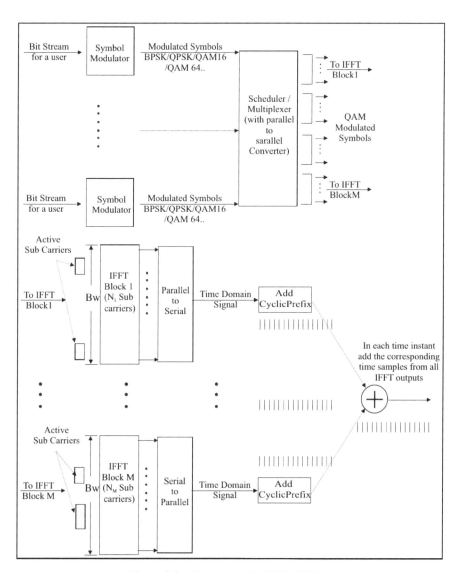

Figure 2.8 Transmitter for VSB OFDM.

a system whose baseband transmitter architecture at the base station (BS) is given in Figure 2.8. The system of the ASB OFDM is also referred to as VSB OFDM equivalently. The system is assumed to support multiple users simultaneously across different subcarriers. There are different groups

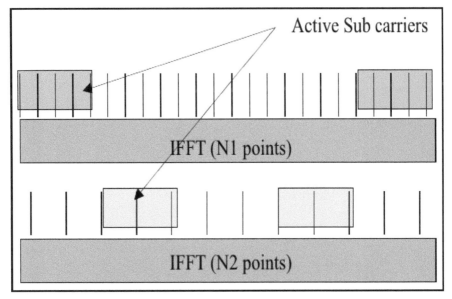

Figure 2.9   Frequency domain configuration of VSB OFDM.

of subcarriers, where the groups have different subcarrier bandwidths. Each IFFT spans the entire system bandwidth by using the same sampling period. In each IFFT, N1, N2, N3, etc., denote the number of subcarriers, where larger numbers generate narrower subcarrier bandwidths. It can be noted that only a fraction of the subcarriers in each IFFT is activated (the number of active subcarriers in each IFFT can be made to vary), and that the active subcarriers of the different groups are selected such that the frequency bands spanned by the different groups do not overlap. The frequency domain view of such a configuration is given in Figure 2.9. A user must be allocated to a particular subcarrier group, whose subcarrier bandwidth suits the requirement of the user conditions optimally. At the receiver, a flexible FFT can be implemented as a user will need only one type of subcarrier at a time, while at the BS as many IFFTs are used as subcarrier types. It can be seen that since sub-bands are next to each other, and since the symbol durations are not the same, the orthogonality will no longer be maintained. Therefore, the system can also be called non-OFDM. However, not all subcarriers will be non-orthogonal. Some subcarriers whose frequencies are an integer multiple of another will still be orthogonal. The receiver at the BS can be as in Figure 2.10.

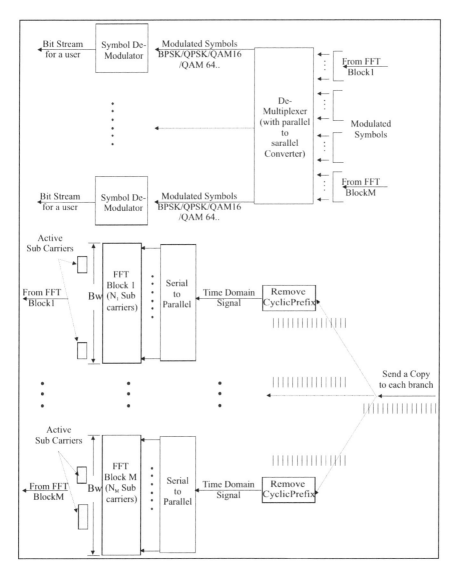

Figure 2.10   Receiver for VSB OFDM.

2.4 Windowed OFDM

The OFDM signal produces large side lobes in the spectrum due to the rectangular pulse shape in the time domain signal. In windowed CP-OFDM technique time domain samples of CP-OFDM pulse are windowed to suppress the symbol energy at the edge of the CP-OFDM symbol [WHKJ04]. Windowing transmitted OFDM symbols allow the amplitude to go smoothly to zero at the symbol boundaries leading to reduced discontinuity between symbols in time. This induces the spectrum of the transmitted signal to go down more rapidly. To maintain orthogonality among subcarriers' additional guard samples are cyclically inserted. However, additional extended symbols decrease the spectrum efficiency.

2.4.1 Transmitter

We begin by considering the OFDM-CP signal, $\mathbf{x}_{CP}$, given in (2.2). We consider L_G to be the length of additional extended symbols. Last L_G samples of $\mathbf{x}$ are prefixed at the starting of the samples, $\mathbf{x}_{CP}$. Also, first L_G samples of $\mathbf{x}$ are postfixed at the end of the samples, $\mathbf{x}_{CP}$. Extended vector of length, $N + L_{CP} + 2L_W$, $\mathbf{x}_e$ can be given as $\mathbf{x}_e = \mathbf{\Phi x}$, where $\mathbf{\Phi}$ can be given as,

$$
\mathbf{\Phi} = \left[\begin{array}{ccc} \mathbf{0}_{L_{CP} \times (N - L_{CP} - L_G)} & | & \mathbf{I}_{L_{CP} + L_G} \\ & \mathbf{I}_N & \\ \mathbf{I}_{L_G}| & \mathbf{0}_{L_{CP} \times (N - L_G)} & \end{array} \right]_{(N + L_{CP} + 2L_G) \times N}
$$

(2.8)

Further, the extended symbols are smoothed using windowing. Let $\mathbf{g}$ be the L_G length row vector which holds the coefficients of pulse shape which is used to smooth first L_G samples of $\mathbf{x}_e$. The transmitted signal can be $\mathbf{x}_t = \mathbf{w} \circ \mathbf{x}_e$, where $\mathbf{w} = [\mathbf{g} \ \mathbf{1}_{N+L_{CP}} \ \mathbf{1}_{L_G} - \mathbf{g}]^T$, $\mathbf{1}_-$ is a $(-)$ length row vector whose all entries are 1s and $\circ$ represents Hadamard point-to-point vector multiplication. Adjacent symbols are overlapped in extended symbol duration to reduce the spectral efficiency loss. The spectral efficiency for windowed OFDM is $\frac{N}{N+L_{CP}+L_G}$ whereas for CP-OFDM, the spectral efficiency is $\frac{N}{N+L_{CP}}$. The transmitter processing can be understood in the light of Figure 2.11.

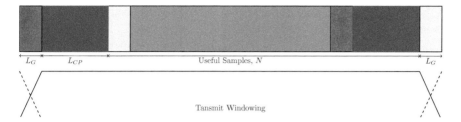

Figure 2.11 Windowed OFDM processing: transmitter side.

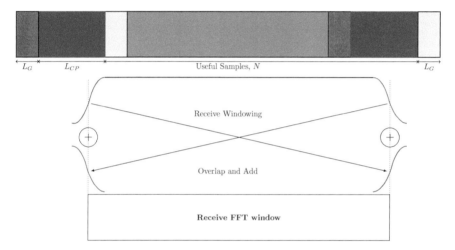

Figure 2.12 Windowed OFDM processing: receiver side.

2.4.2 Receiver

At the receiver, symbols in extended period are discarded to eliminate the effect of windowing at the transmitter. Further CP symbols can be discarded and the signal can be written in a similar way as in (2.4). Further receiver processing can be similar as in Section 2.2.2.

In case of multi-user interference (MUI), windowing at the receiver helps in reducing the MUI [ZMSR16, MZSR17]. Figure 2.12 explains the receiver signal processing steps. In case of windowing at the receiver, CP is not discarded at the first; however, symbols in the extended period are discarded and signal is $\mathbf{y}_{cp}$ as in (2.3). $\mathbf{y}_{cp}$ is first windowed using a window having smooth edges to reduce multi-user interference [ZMSR16, MZSR17]. We consider a window of length $N + 2L_{rx}$, where $2L_{rx}$ is the length of smooth pulse shape at each edge. We define K_o to be the starting sample index of windowing; we may call it window offset, $K_o \leq L_{cp} - 2L_{rx}$. Let $\mathbf{g}_r$ be a $2L_{rx}$ length row vector which holds pulse shape coefficients which is used to smooth first few samples of $\mathbf{y}_{cp}$. The $N + 2L_{rx}$ length windowed vector can be given as $\mathbf{y}_w = \mathbf{w}_{rx} \circ \mathbf{y}_{cp}$, where $\mathbf{w}_{rx} = [\mathbf{0}_{K_o} \ \mathbf{g} \ \mathbf{1}_{N-2L_{rx}} \ \mathbf{1}_{2L_{rx}} - \mathbf{g} \ \mathbf{0}_{L_{CP}-K_o-2L_{rx}}]^{\mathrm{T}}$, where $\mathbf{0}_-$ is a $(-)$ length row vector whose all entries are 0s. It is evident that first and last $2L_{rx}$ samples of $\mathbf{y}_w$ are distorted due to windowing. Interestingly, samples of $\mathbf{y}_w$ are cyclic in nature. We exploit this property to get useful N samples, which can be recovered from $N + 2L_{rx}$ samples of $\mathbf{y}_w$ by using the principle of overlap and add. To do so, the first $2L_{rx}$ samples are added to the last $2L_{rx}$ samples. We collect first $2L_{rx}$ samples in a row vector, $\mathbf{r} = [\mathbf{y}_w((N-1-L_{CP}+K_o+L_{rx}+1) \mod N) : \mathbf{y}_w((N-1-L_{CP}+K_o+2L_{rx}) \mod N)] \circ \mathbf{g}_{rx}$, and last $2L_{rx}$ samples in a row vector, $\mathbf{s} = [\mathbf{y}_w((N-1-L_{CP}+K_o+L_{rx}+1) \mod N) : \mathbf{y}_w((N-1-L_{CP}+K_o+2L_{rx}) \mod N)] \circ (\mathbf{1} - \mathbf{g}_{rx})$. Overlap and added vector, $\mathbf{t} = (\mathbf{r}+\mathbf{s})^{\mathrm{T}} = [\mathbf{y}_w((N-1-L_{CP}+K_o+L_{rx}+1) \mod N) : \mathbf{y}_w((N-1-L_{CP}+K_o+2L_{rx}) \mod N)]$. Thus, the effect of receiver windowing is eliminated in s. Vector s replaces first and last $2L_{rx}$ samples. Further, first and last L_{rx} samples are discarded to obtain N useful data symbols. The obtained signal can be written in a similar way as in (2.4). Further receiver processing can be similar as in Section 2.2.2.

2.5 Filtered OFDM

In filtered-OFDM (F-OFDM) [AJM15, ZIX+17], the available band is divided into multiple subbands. Each sub-band may use different OFDM parameters optimized for the application such as frequency spacing, cyclic prefix and transmit time interval. Each sub-band spectrum is filtered to avoid inter-sub-band interference. The main advantage is that different users (sub-bands) do not need to be time synchronized; as a result, global synchronization is relaxed. F-OFDM can achieve desirable frequency localization while enjoying the benefits of CP-OFDM and designing the filter appropriately. F-OFDM allows co-existence of different signal components with different OFDM primitives. By enabling multiple parameter configurations, F-OFDM is able to provide more optimum parameter numerology choice for each service group and hence better overall system efficiency. The impulse response of an ideal LPF is a sinc function which is infinitely long. Hence, for practical implementation, the sinc function is soft truncated with different window functions (1) Hann window (2) root-raised cosine (RRC)

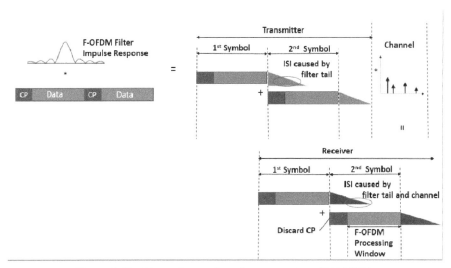

Figure 2.13 Transmitter and receiver processing of F-OFDM.

window. In this way, the impulse response of the obtained filters will fade out quickly, thus limiting ISI introduced between consecutive OFDM symbols. Transmitter and receiver processing is demonstrated in Figure 2.13.

2.5.1 Transmitter

Let us consider an F-OFDM system with N subcarriers that are divided into K sub-bands with each sub-band having M subcarriers, i.e., $N = MK$. Transmitted symbols can be given as,

$$\mathbf{d} = [\mathbf{d}_1 \ \mathbf{d}_2 \cdots \mathbf{d}_K]^{\mathrm{T}}, \tag{2.9}$$

where

$$\mathbf{d}_k = [d_k(1) \ d_k(1) \cdots d_M]^{\mathrm{T}}, \tag{2.10}$$

is the transmitted signal for the kth sub-band, $1 \leq k \leq K$. The transmitted signal before filtering can be given as $\mathbf{x}'_k = \mathbf{\Phi}'\mathbf{W}_k\mathbf{d}_k$, where $\mathbf{\Phi}'$ is the CP addition matrix which can be obtained by putting $L_G = 0$ in (2.8) and $\mathbf{W}_k$ is an $N \times M$ matrix which is obtained selecting $[(k-1)M + 1]^{\mathrm{th}}$ to $(kM)^{\mathrm{th}}$ columns of $\mathbf{W}_N$. Let $\mathbf{g}_k$ be the kth sub-band filter impulse response, which can be given as, $\mathbf{g}_k = [g_k(0) \ g_k(1) \cdots g_k(L_F - 1)]$, where L_F is the filter length. Let us define a $(N + L_{CP} + L_F - 1) \times N + L_{CP}$ size teoplitz matrix, $\mathbf{G}_k$

whose first column is $[\mathbf{g}_k \ \mathbf{0}_{N+L_{CP}-1}]^{\mathrm{T}}$ and first row is $[g_k(0) \ \mathbf{0}_{N+L_{CP}-1}]^{\mathrm{T}}$. The transmitted signal for the kth sub-band can be given as,

$$\mathbf{x}_k = \mathbf{G}_k \mathbf{\Phi}' \mathbf{W}_k \mathbf{d}_k. \tag{2.11}$$

It is evident that the transmitted signal is spread in time domain due to filtering as the transmitted signal is $N + L_{CP} + L_F - 1$ long. As it is shown in Figure 2.13, the current symbol at the kth sub-band will overlap with the next and previous symbols. Thus, ISI is introduced due to filter tail.

2.5.2 Receiver Processing

Let $\mathbf{h}_k = [h(0), \ h_k(1), \cdots h_k(L-1)]^{\mathrm{T}}$ be L length channel impulse response vector for the kth subband. Received vector of length $N+L_{CP}+L_F+L-2$ is given by,

$$\mathbf{y}_k = \mathbf{h}_k * \mathbf{x}_k + \boldsymbol{\nu}_k, \tag{2.12}$$

where $\boldsymbol{\nu}_k$ is the AWGN vector of length $N+L_{CP}+L_F+L-2$ with elemental variance σ_ν^2. Let us define a $(N + L_{CP} + L_F + L - 2) \times (N + L_{CP} + L_F - 1)$ teoplitz matrix, $\mathbf{H}_k'$ whose first row is $[h_k(0) \ \mathbf{0}_{N+L_{CP}+L_F-2}]$ and first column is $[\mathbf{h}_k \ \mathbf{0}_{N+L_{CP}+L_F-2}]^{\mathrm{T}}$. Let K_o be the offset in receiver windowing. $N+L_{CP}$ symbols are selected from $\mathbf{y}_k$ as $\mathbf{z}_k = \mathbf{y}_k[K_o : K_o+N+L_{CP}-1]$. Let $\mathbf{H}_k$ be formed by selecting $(K_o+1)^{\text{th}}$ to $(K_o+N+L_CP)^{\text{th}}$ row. Signal $\mathbf{z}_k$ can be given as,

$$\mathbf{z}_k = \mathbf{H}_k \mathbf{x}_k + \tilde{\boldsymbol{\nu}}, \tag{2.13}$$

where $\tilde{\boldsymbol{\nu}}$ is the $N + L_{cp}$ length noise vector. Next, $\mathbf{z}_k$ is passed through linear filtering, CP removal, and DFT processing one by one. The processed signal can be given as,

$$\mathbf{z}_k^I = \mathbf{W}_k^{\mathrm{H}} \mathcal{T} \mathbf{G}_k^{\mathrm{H}} \mathbf{z}_k, \tag{2.14}$$

where $\mathcal{T} = [\mathbf{0}_{N \times L_{CP}} \ \mathbf{I}_N]$ is the CP removal matrix. $\mathbf{z}_k^I$ suffers from ISI due to filter tail and channel dispersion. Let $\boldsymbol{\phi} = \mathbf{W}_k^{\mathrm{H}} \mathcal{T} \mathbf{G}_k^{\mathrm{H}} \mathbf{G}_k \mathbf{\Phi}' \mathbf{W}_k$. A ZF receiver can be employed to combat ISI. The ZF equalized signal can be given as,

$$\mathbf{z}_k^{zf} = (\boldsymbol{\phi}^{\mathrm{H}} \boldsymbol{\phi})^{-1} \boldsymbol{\phi}^{\mathrm{H}} \mathbf{z}_k^I. \tag{2.15}$$

2.6 GFDM

GFDM is a multi-carrier modulation technique with some similarity to OFDM. We begin by considering a block of QAM modulated symbols

$\mathbf{d} = [d_0 \ d_1 \cdots d_{MN-1}]^{\mathrm{T}}$. We assume that data symbols are independent and identical, i.e., $E[d_l d_l^*] = \sigma_d^2$, $\forall l$ and $E[d_l d_q^*] = 0$ when $l \neq q$. Let the total bandwidth B be divided into N number of subcarriers where the symbol duration $T = \frac{N}{B}$ second and $\frac{B}{N}$ Hz is the subcarrier bandwidth. In case of OFDM, this leads to orthogonal subcarriers. Let the symbol duration, T, be one time slot. GFDM is a block-based transmission scheme and we consider a block to have M such time slots. Hence, in one block, there are N subcarriers $\times$ M timeslots $= NM$ QAM symbols.

2.6.1 Transmitter

The $MN \times 1$ data vector $\mathbf{d} = [d_{0,0} \cdots d_{k,m} \cdots d_{N-1,M-1}]^{\mathrm{T}}$, where $k = 0 \cdots N - 1$ denotes subcarrier index and $m = 0 \cdots M - 1$ indicates the time slot index. The data vector, $\mathbf{d}$, is modulated using a GFDM modulator. Data $d_{k,m}$ are first upsampled by N, which are represented as,

$$d_{k,m}^{up}(n) = d_{k,m}\delta(n - mN), \ n = 0, 1, \cdots, MN - 1. \qquad (2.16)$$

Now, these upsampled data are pulse shaped. The impulse response of the pulse shaping filter is represented by $g(n)$. Its length is MN. The upsampled data $d_{k,m(n)}^{up}$ are circularly convoluted with such a pulse shaping filter $g(n)$ and can be written as,

$$x_{k,m}^{f}(n) = \sum_{r=0}^{MN-1} d_{k,m}g(n - r)_{MN}\delta(r - mN) = d_{k,m}g(n - mN)_{MN}. \qquad (2.17)$$

Now the filtered data are up-converted to the kth subcarrier frequency and is given as,

$$x_{k,m}(n) = d_{k,m}g(n - mN)_{MN}e^{\frac{j2\pi kn}{N}} = d_{k,m}a_{k,m}(n), \qquad (2.18)$$

where $a_{k,m}(n) = g(n - mN)_{MN}e^{\frac{j2\pi kn}{N}}$ for $n = 0, 1, \cdots MN - 1$ is called the kernel of GFDM for the kth subcarrier and m^{th} time slot. Now, output samples of all subcarrier frequencies and time slots are added and the GFDM signal can be given as,

$$x(n) = \sum_{m=0}^{M-1}\sum_{k=0}^{N-1} x_{k,m}(n) = \sum_{m=0}^{M-1}\sum_{k=0}^{N-1} d_{k,m}a_{k,m}(n). \qquad (2.19)$$

If we let $l = mN + k$ for $m = 0, 1 \cdots M - 1$ and $k = 0, 1 \cdots N - 1$, then, we may write (2.19) as,

$$x(n) = \sum_{l=0}^{MN-1} a_l^N(n) d_l^N, \qquad (2.20)$$

where superscript N is the identifier of the specific mapping between index l and time slot and subcarrier index tuple (k, m) which takes N as scaler multiplier in the mapping. If we collect all output samples of the GFDM signal in a vector called $\mathbf{x} = [x(0)\ x(1) \cdots x(MN - 1)]^{\mathrm{T}}$ and all samples of $a_l^N(n)$ in a vector called $\mathbf{a}_l^N = [a_l^N(0)\ a_l^N(1) \cdots a_l^N(MN - 1)]$, then using (2.20), $\mathbf{x}$ can be written as,

$$\mathbf{x} = \sum_{l=0}^{MN-1} \mathbf{a}_l^N d_l^N \equiv \sum_{l=0}^{MN-1} d_l^N \times l^{\mathrm{th}} \text{ column vector } \mathbf{a}_l^N = \mathbf{A} \mathbf{d}^N,$$

$$(2.21)$$

where $\mathbf{A} = [\mathbf{a}_0^N\ \mathbf{a}_1^N \cdots \mathbf{a}_{MN-1}^N]$ is the modulation matrix and $\mathbf{d}^N = [d_0\ d_1 \cdots d_l \cdots d_{NM-1}]^{\mathrm{T}}$ is the precoded data vector. If we collect all samples of $a_{k,m}(n)$ in a vector called $\mathbf{a}_{k,m} = [a_{k,m}(0)\ a_{k,m}(0) \cdots a_{k,m}(MN - 1)]^{\mathrm{T}}$, then $\mathbf{A}$ can also be written as,

$$
\mathbf{A} = [\underbrace{\overbrace{\mathbf{a}_{0,0}}^{1^{st}\ \text{freq.}}\ \overbrace{\mathbf{a}_{1,0}}^{2^{nd}\ \text{freq.}} \cdots \overbrace{\mathbf{a}_{N-1,0}}^{N^{th}\ \text{freq.}}}_{1^{st}\text{time slot}} | \underbrace{\mathbf{a}_{0,1}\ \mathbf{a}_{1,1} \cdots \mathbf{a}_{N-1,1}}_{2^{nd}\text{time slot}} |
$$

$$
\cdots | \underbrace{\mathbf{a}_{0,M-1}\ \mathbf{a}_{1,M-1} \cdots \mathbf{a}_{N-1,M-1}}_{M^{th}\text{time slot}}]
$$

$$
= \begin{bmatrix}
\underbrace{g(0)}_{\mathbf{a}_{0,0}} & \underbrace{g(0)}_{\mathbf{a}_{1,0}} & \cdots & \cdots & \underbrace{g(0)}_{\mathbf{a}_{N-1,0}} & \underbrace{g(MN\text{-}N)}_{\mathbf{a}_{0,1}} & \cdots & \cdots & \underbrace{g(MN\text{-}N)}_{\mathbf{a}_{N-1,1}} & \cdots \\
g(1) & g(1)\rho & \cdots & \cdots & g(1)\rho^{N\text{-}1} & g(MN\text{-}N\text{+}1) & \cdots & \cdots & g(MN\text{-}N\text{+}1)\rho^{N\text{-}1} & \vdots \\
\vdots & \vdots & \cdots & \cdots & \vdots & \vdots & \cdots & \cdots & & (M\text{-}2)N \text{ terms} \\
\vdots & \vdots & \cdots & \cdots & \vdots & \vdots & \cdots & \cdots & \cdots & \vdots \\
g(MN\text{-}1) & g(MN\text{-}1)\rho^{MN\text{-}1} & \cdots & \cdots & g(MN\text{-}1)\rho^{(N\text{-}1)(MN\text{-}1)} & g(MN\text{-}N\text{-}1) & \cdots & \cdots & g(MN\text{-}N\text{-}1)\rho^{(N\text{-}1)(MN\text{-}1)} & \cdots
\end{bmatrix},
$$

$$(2.22)$$

where $\rho = e^{\frac{j2\pi}{N}}$. At this point, it is also interesting to look into the structure $\mathbf{a}_{k,m}$ vectors which constitute columns of $\mathbf{A}_N$. The first column of $\mathbf{A}_N$, i.e., $\mathbf{a}_{0,0}$ holds all coefficients of the pulse shaping filter and other columns of $\mathbf{A}_N$ or other vectors in the set of $\mathbf{a}_{k,m}$'s are time and frequency shifted version of $\mathbf{a}_{0,0}$ where the frequency index k denotes $\frac{k}{N}$ shift in frequency and time index m denotes m time slot or mN sample cyclic shift. Taking clue from the description of column vectors $\mathbf{a}_{k,m}$, the structure of $\mathbf{A}_N$ can be understood by (2.22). Columns having the same time shift are put together. Columns which are N column index apart are time shifted version of each other. Columns having the same time shift are arranged in increasing order of frequency shift. We will explore time domain and frequency domain behavior for an example $\mathbf{A}_N$ with total subcarriers $N = 4$ and total time slots $M = 5$. The pulse shaping filter is taken to be RRC with a roll-off factor (ROF) of 0.9. Figure 2.14 shows the absolute values of each column index $u = 0, 1 \cdots MN - 1$ with sample index $n = 0, 1 \cdots, MN - 1$. It can be observed that first $N = 4$ columns have 0 time shift and next N columns have unit time shift, and so on. Further, Figure 2.15 provides the time–frequency view of $\mathbf{A}$.

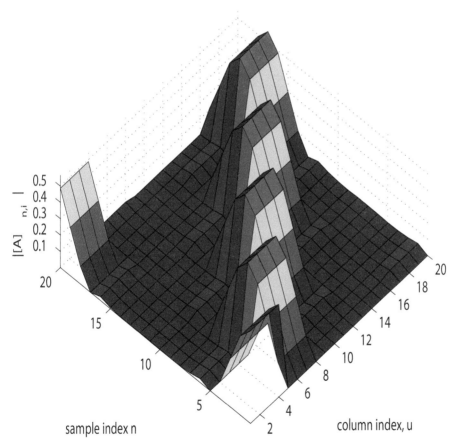

Figure 2.14 Time domain view of columns of $\mathbf{A}$ for $N = 4$, $M = 5$, and $ROF = 0.9$.

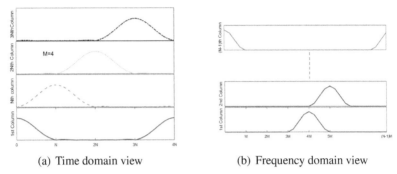

(a) Time domain view   (b) Frequency domain view

Figure 2.15   Time–frequency view of GFDM matrix $\mathbf{A}$. $M = 5$.

Alternatively, if we take $l = kM + m$ in (2.19), modulation matrix $\mathbf{A}_M$ can be represented as,

$$
\mathbf{A}_M = \Big[\underbrace{\overbrace{\mathbf{a}_{0,0}}^{1^{st}\text{ time}} \overbrace{\mathbf{a}_{0,1}}^{2^{nd}\text{ time}} \cdots \overbrace{\mathbf{a}_{0,M-1}}^{M^{th}\text{ time}}}_{1^{st}\text{frequency}} \Big| \underbrace{\mathbf{a}_{1,0}\ \mathbf{a}_{1,1} \cdots \mathbf{a}_{1,M-1}}_{2^{nd}\text{frequency}} \Big|
$$
$$
\cdots \Big| \underbrace{\mathbf{a}_{N-1,0}\ \mathbf{a}_{N-1,1} \cdots \mathbf{a}_{N-1,M-1}}_{N^{th}\text{frequency}} \Big]. \tag{2.23}
$$

Matrix $\mathbf{A}_N$ and $\mathbf{A}_M$ can be related as $\mathbf{A}_N = \zeta \mathbf{A}_M$, where ζ is a permutation matrix which permutes column of matrix applied. Cyclic prefix is added to the GFDM modulated block to prevent inter-block interference in FSFC. CP of length N_{CP} is prepended to $\mathbf{x}$. After adding CP, the transmitted vector, $\mathbf{x}_{cp}$, can be given as

$$
\mathbf{x}_{cp} = [\mathbf{x}(MN - N_{cp} + 1 : MN)\ ;\ \mathbf{x}] \tag{2.24}
$$

In the rest of the chapter, for equations which are valid for both $\mathbf{A}_M$ and $\mathbf{A}_N$, the modulation matrix will be denoted by $\mathbf{A}$.

2.6.2 Receiver

Let $\mathbf{h} = [h_1,\ h_2, \cdots h_L]^{\mathrm{T}}$ be the L length channel impulse response vector, where h_r, for $1 \leq r \leq L$, represents the complex baseband channel coefficient of the rth path [PS08], which we assume is ZMCSC. We also assume that channel coefficients related to different paths are uncorrelated.

We consider $N_{cp} \geq L$. The received vector of length $N_{CP} + NM + L - 1$ is given by,

$$\mathbf{y}_{cp} = \mathbf{h} * \mathbf{x}_{cp} + \boldsymbol{\nu}_{cp}, \tag{2.25}$$

where $\boldsymbol{\nu}_{cp}$ is the AWGN vector of length $MN + N_{cp} + L - 1$ with elemental variance σ_ν^2. The first N_{cp} samples and last $L - 1$ samples of $\mathbf{y}_{cp}$ are removed at the receiver, i.e., $\mathbf{y} = [\mathbf{y}_{cp}(N_{cp} + 1 : N_{cp} + MN)]$. Use of cyclic prefix converts linear channel convolution to circular channel convolution when $N_{cp} \geq L$ [STB11]. The MN length received vector after the removal of CP can be written as,

$$\mathbf{y} = \mathbf{HAd} + \boldsymbol{\nu}, \tag{2.26}$$

where $\mathbf{H}$ is circulant convolution matrix of size $MN \times MN$, which can be written as,

$$\mathbf{H} = \begin{bmatrix} h_1 & 0 & \cdots & 0 & h_L & \cdots & h_2 \\ h_2 & h_1 & \cdots & 0 & 0 & \cdots & h_3 \\ \vdots & & \ddots & & & & \\ h_L & h_{L-1} & \cdots & \cdots & \cdots & \cdots & 0 \\ 0 & h_L & \cdots & \cdots & \cdots & \cdots & 0 \\ \vdots & & \ddots & & & & \\ 0 & 0 & & h_L & \cdots & \cdots & h_1 \end{bmatrix}, \tag{2.27}$$

and $\boldsymbol{\nu}$ is the WGN vector of length MN with elemental variance σ_ν^2. Received vector $\mathbf{y}$ is distorted due to (i) self-interference as subcarriers are non-orthogonal [MKLFda] and (ii) inter-carrier interference(ICI) due to FSFC.

To equalize the channel and GFDM-induced self-interference, a joint MMSE equalizer [MMG$^+$14] is considered here. Equalized data symbol vector, $\hat{\mathbf{d}}_{JP}$, can be given as, $\hat{\mathbf{d}}_{JP} = \mathbf{B}_{eq}\mathbf{y}$, where $\mathbf{B}_{eq}$ is the joint-MMSE equalizer matrix and can be given as,

$$\mathbf{B}_{eq} = \boldsymbol{\Theta}^{-1}[\tfrac{\sigma_\nu^2}{\sigma_d^2}\mathbf{I} + (\mathbf{HA})^{\mathrm{H}}\mathbf{HA}]^{-1}(\mathbf{HA})^{\mathrm{H}} \tag{2.28}$$

where $\boldsymbol{\Theta}^{-1}$ is the diagonal bias correction matrix for joint-processing, where,

$$\boldsymbol{\Theta} = diag\{[\tfrac{\sigma_\nu^2}{\sigma_d^2}\mathbf{I} + (\mathbf{HA})^{\mathrm{H}}\mathbf{HA}]^{-1}(\mathbf{HA})^{\mathrm{H}}\mathbf{HA}\}$$

.

2.7 Precoded GFDM

In this section, precoding schemes to enhance the performance of GFDM are described. Let P be a precoding matrix of size $MN \times MN$. The data vector d be multiplied with precoding matrix P and we obtain precoded data vector $\tilde{d} = Pd$. Conventional GFDM system can be seen as a special case of precoded GFDM system when $P = I_{NM}$. The precoded data vector, $\tilde{d}$, is modulated using GFDM modulator as described in Section 2.6.1. The first precoding scheme BIDFT is described in Section 2.7.1. The scheme, is developed using special properties of $(\mathbf{HA})^H \mathbf{HA}$ and $\mathbf{A}^H \mathbf{A}$. For this precoding scheme, there can be two kinds of receiver processing, namely, (i) joint processing: described in Section 2.7.1, which equalizes the channel and the GFDM modulation matrix simultaneously, whereas (ii) two-stage processing: described in Section 2.7.1, first stage equalizes for the channel and second stage equalizes for the GFDM modulation matrix. Two types of precoding matrices are defined for BIDFT precoding, namely, (i) BIDFT-M: where $\mathbf{A}$ is structured in blocks of $M \times M$, i.e. $\mathbf{A}_M$ and (ii) BIDFT-N: where $\mathbf{A}$ is structured in blocks of $N \times N$, i.e. $\mathbf{A}_N$. BIDFT-N precoded GFDM is processed using joint processing as well as two-stage processing, whereas BIDFT-M precoded GFDM is processed using two-stage processing only.

DFT-based precoding is described in Section 2.7.2. SVD-based precoding is described in Section 2.7.3. For each precoding scheme, precoder matrix $\mathbf{P}$, corresponding receivers, and post-processing SNR are described in detail. Both BIDFT and DFT-based precoding does not require channel state information (CSI) at the transmitter to compute $\mathbf{P}$, whereas SVD-based precoding needs CSI at the transmitter to compute $\mathbf{P}$. Channel knowledge at the transmitter can be maintained via feedback from the receiver or through the reciprocity principle in a duplex system [CG01].

2.7.1 Block IDFT Precoded GFDM

The received signal in (2.26) can be processed in two ways, (i) joint processing: channel and self-interference are equalized simultaneously and (ii) two-stage processing: channel and self-interference and equalized separately.

2.7.1.1 Joint processing

Suppose the received signal in (2.26) passed through a matched filter. The equalized vector can be given as,

$$
\begin{aligned}
\mathbf{y}^{\mathrm{MF}} &= (\mathbf{HAP})^H \mathbf{y} \\
&= \mathbf{P}^H (\mathbf{HA})^H \mathbf{HAPd} + (\mathbf{HA})^H \boldsymbol{\nu}.
\end{aligned}
\tag{2.29}
$$

$(\mathbf{HA}_N)^H\mathbf{HA}_N$ is a block circulant matrix with blocks of size $N \times N$ (it may be noted here that $(\mathbf{HA}_M)^H\mathbf{HA}_M$ will not be block circulant with blocks of size $M \times M$) [TDB15]. Since $(\mathbf{HA}_N)^H\mathbf{HA}_N$ is a block circulant matrix with blocks of size $N \times N$, it can be decomposed as given in [Tra73, DG83], as $(\mathbf{HA}_N)^H\mathbf{HA}_N = \mathbf{F}_{bN}\mathbf{D}_{bN}\mathbf{F}_{bN}^H$, where $\mathbf{F}_{bN} = \begin{bmatrix} \mathbf{W}_N^0 & \mathbf{W}_N & \cdots \mathbf{W}_N^{M-1} \end{bmatrix}_{MN \times MN}$, where $\mathbf{W}_N^i = \dfrac{\begin{bmatrix} \mathbf{I}_N & w_N^i\mathbf{I}_N & \cdots & w_N^{i(N-1)}\mathbf{I}_N \end{bmatrix}^T_{N \times MN}}{\sqrt{N}}$, where $w_N = e^{\frac{j2\pi}{N}}$ and $\mathbf{D}_{bN} = diag\{\mathbf{D}_{bN}^0 \ \mathbf{D}_{bN}^1 \cdots \mathbf{D}_{bN}^{M-1}\}$ is block diagonal matrix with blocks of size $N \times N$, where $\mathbf{D}_{bN}^r$ is the rth diagonal matrix of size $N \times N$.

Using this decomposition and taking $\mathbf{A}$ as $\mathbf{A}_N$, the matched filter output in (2.29) can be written as,

$$\mathbf{y}^{MF} = \mathbf{P}^H\mathbf{F}_{bN}\mathbf{D}_{bN}\mathbf{F}_{bN}^H\mathbf{Pd} + (\mathbf{HA}_N)^H\boldsymbol{\nu}. \tag{2.30}$$

2.7.1.2 BIDFT-N precoding

If we choose, $\mathbf{P} = \mathbf{F}_{bN}$ (call it block inverse discrete Fourier transform -N (BIDFTN) precoding), then,

$$\mathbf{y}^{MF} = \mathbf{D}_{bN}\mathbf{d} + (\mathbf{HA}_N)^H\boldsymbol{\nu}$$

$$= \begin{bmatrix} \mathbf{D}_{bN}^0 & & & \\ & \mathbf{D}_{bN}^1 & & \\ & & \ddots & \\ & & & \mathbf{D}_{bN}^{N-1} \end{bmatrix} \begin{bmatrix} \mathbf{d}_0^N \\ \mathbf{d}_1^N \\ \vdots \\ \mathbf{d}_{M-1}^N \end{bmatrix} + \bar{\boldsymbol{\nu}}, \tag{2.31}$$

where $\bar{\boldsymbol{\nu}}$ is MF processed noise vector. In the above equation $\mathbf{D}_{bN}$ being block diagonal matrix, adds only $N-1$ interfering symbols instead $MN-1$ (in case of uncoded GFDM). This shows that precoding reduces the number of interfering symbols significantly. ZF equalization is applied to reduce interference further. Multiplying $\mathbf{D}_{bN}^{-1}$ in the above equation, we get,

$$\hat{\mathbf{d}}_{bidft}^{JP} = \mathbf{d} + \boldsymbol{\nu}_{bdft}^{JP}, \tag{2.32}$$

where $\boldsymbol{\nu}_{bdft}^{JP} = \mathbf{D}_{bN}^{-1}(\mathbf{HA}_N)^H\boldsymbol{\nu}$ is post-processing noise vector and superscript JP signifies that signal processing steps followed in this method are joint processing (channel and self-interference are equalized jointly). $\mathbf{D}_{bN}$ can be computed as,

$$\mathbf{D}_{bN} = \mathbf{F}_{bN}^H(\mathbf{HA}_N)^H\mathbf{HA}_N\mathbf{F}_{bN}. \tag{2.33}$$

Post-processing SNR for the lth symbol can be obtained as,

$$\gamma_{bidft,l}^{JP} = \frac{\sigma_{d^2}}{E[\boldsymbol{\nu}_{bdft}^{JP}(\boldsymbol{\nu}_{bdft}^{JP})^{\text{H}}]_{l.l}}, \tag{2.34}$$

where the denominator in the above equation is the post-processing noise power for the lth symbol.

2.7.1.3 Two-stage processing

In the above method, channel and GFDM were equalized together. In this method, we will first equalize channel distortions, and then GFDM induced self-interference. As explained earlier, $\mathbf{H}$ is a circulant matrix. Hence, $\mathbf{H}$ can be decomposed as,

$$\mathbf{H} = \mathbf{W}_{NM}\boldsymbol{\Psi}\mathbf{W}_{NM}^{\text{H}}, \tag{2.35}$$

where $\mathbf{W}_{NM}$ is the normalized IDFT matrix of size $MN \times MN$ and $\boldsymbol{\Psi} = diag\{v_0, v_1, \cdots v_{MN-1}\}$ is a diagonal matrix. The channel equalized vector can be obtained as,

$$\mathbf{y}_{FDE} = \mathbf{W}_{NM}^{\text{H}}\boldsymbol{\Psi}^{-1}\mathbf{W}_{NM}\mathbf{y} = \mathbf{AP}\mathbf{d} + \mathbf{W}_{NM}\boldsymbol{\Psi}^{-1}\mathbf{W}_{NM}^{\text{H}}\boldsymbol{\nu}, \tag{2.36}$$

where the first term is the transmitted signal which is free from channel distortions completely, the second term is enhanced noise, and subscript FDE is the acronym for frequency domain equalization (as above described channel equalization is frequency domain equalization [STB11]). $\boldsymbol{\Psi}$ can be obtained as $\boldsymbol{\Psi} = \mathbf{W}_{NM}^{\text{H}}\mathbf{H}\mathbf{W}_{NM}$ which is equivalently obtained by taking NM point FFT of zero-padded channel convolution vector $\mathbf{h}$ which is also the first column of $\mathbf{H}$ [STB11, HT10].

Now passing channel equalized data $\mathbf{y}_{FDE}$ to matched filter receiver, we can get,

$$\mathbf{y}_{FDE-MF} = (\mathbf{AP})^{\text{H}}\mathbf{y}_{FDE} = \mathbf{P}^{\text{H}}\mathbf{A}^{\text{H}}\mathbf{AP}\mathbf{d} + (\mathbf{AP})^{\text{H}}\mathbf{W}_{NM}\boldsymbol{\Psi}^{-1}\mathbf{W}_{NM}^{\text{H}}\boldsymbol{\nu}. \tag{2.37}$$

It is shown in [TDB15] that $\mathbf{A}^{\text{H}}\mathbf{A}$ is BCCB matrix with blocks of size either $N \times N$ or $M \times M$, which depends on whether $\mathbf{A} = \mathbf{A}_N$ or $\mathbf{A} = \mathbf{A}_M$. An example of $\mathbf{A}^{\text{H}}\mathbf{A}$ matrix for $N = 4$, $M = 3$ is shown in Figure 2.16.

2.7.1.4 BIDFT-N precoding

When the modulation matrix is defined as $\mathbf{A}_N$, $\mathbf{A}_N^{\text{H}}\mathbf{A}_N$ can be decomposed as,

$$\mathbf{A}_N^{\text{H}}\mathbf{A}_N = \mathbf{F}_{bN}\tilde{\mathbf{D}}_{bN}\mathbf{F}_{bN}^{\text{H}}, \tag{2.38}$$

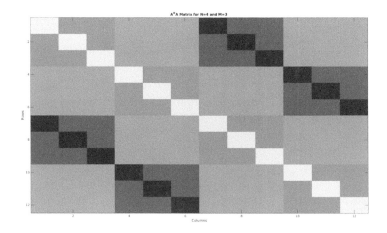

Figure 2.16 An example of $\mathbf{A}^{\text{H}}\mathbf{A}$. It can be seen that blocks of size $M \times M$ are circulant.

where $\tilde{\mathbf{D}}_{bN} = diag\{\mathbf{D}_{bN}^0, \mathbf{D}_{bN}^1 \cdots \mathbf{D}_{bN}^{M-1}\}$ is $MN \times MN$ block diagonal matrix where $\mathbf{D}_{bN}^r$ is r^{th} diagonal block of size $N \times N$. Choosing $\mathbf{P} = \mathbf{F}_{bN}$ (BIDFTN precoding) and using the above decomposition, the matched filter output in (2.37) can be written as,

$$\mathbf{y}_{FDE-MF} = \tilde{\mathbf{D}}_{bN}\mathbf{d} + (\mathbf{AF}_{bN})^{\text{H}}\mathbf{W}_{NM}\mathbf{\Psi}^{-1}\mathbf{W}_{NM}^{\text{H}}\boldsymbol{\nu}. \qquad (2.39)$$

Now multiplying $\tilde{\mathbf{D}}_{bN}^{-1}$ in the above equation,

$$\hat{\mathbf{d}}_{FDE-MF-ZF}^N = \mathbf{d} + \boldsymbol{\nu}_{FDE-MF-ZF}^N, \qquad (2.40)$$

where $\boldsymbol{\nu}_{FDE-MF-ZF}^N = \tilde{\mathbf{D}}_{bN}^{-1}(\mathbf{AF}_{bN})^{\text{H}}\mathbf{W}_{NM}\mathbf{\Psi}^{-1}\mathbf{W}_{NM}^{\text{H}}\boldsymbol{\nu}$ is the enhanced noise vector. Since $\tilde{\mathbf{D}}_{bN}$ needs to be computed to obtain $\hat{\mathbf{d}}_{FDE-MF_ZF}^N$, it can be computed as $\tilde{\mathbf{D}}_{bN} = \mathbf{F}_{bN}\mathbf{A}^{\text{H}}\mathbf{AF}_{bN}^{\text{H}}$. Post-processing SNR for the lth symbol can be obtained as,

$$\gamma_{FDE-MF-ZF,l}^N = \frac{\sigma_{d^2}}{E[\boldsymbol{\nu}_{FDE-MF-ZF}^N(\boldsymbol{\nu}_{FDE-MF-ZF}^N)^{\text{H}}]_{l,l}}, \qquad (2.41)$$

where the denominator in the above equation is the enhanced noise power for the lth symbol.

2.7.1.5 BIDFT-M precoding

Now, if the modulation matrix is defined as $\mathbf{A}_M$, $\mathbf{A}_M^{\text{H}}\mathbf{A}_M = \mathbf{F}_{bM}\mathbf{D}_{bM}\mathbf{F}_{bM}^{\text{H}}$, where $\mathbf{F}_{bM} = \begin{bmatrix} \mathbf{W}_M^0 & \mathbf{W}_M & \cdots & \mathbf{W}_M^{N-1} \end{bmatrix}_{MN \times MN}$, where

$$\mathbf{W}_M^r = \frac{\left[\mathbf{I}_M \quad w_M^r \mathbf{I}_M \quad \cdots \quad w_M^{r(M-1)} \mathbf{I}_M\right]_{M \times MN}^T}{\sqrt{M}}, \text{ where } w_M = e^{\frac{j2\pi}{M}} \text{ and}$$

$\mathbf{D}_{bM} = diag\{\mathbf{D}_{bM}^0 \ \mathbf{D}_{bM}^1 \cdots \mathbf{D}_{bM}^{N-1}\}$ is the block diagonal matrix with blocks of size $M \times M$, where $\mathbf{D}_{bM}^r$ is the rth diagonal matrix of size $M \times M$. Using this decomposition, choosing $\mathbf{P} = \mathbf{F}_{bM}$ (block IDFT-M (BIDFTM) precoding) and following the same signal processing steps as in case of $\mathbf{A}_N$, the equalized data vector can be given as,

$$\hat{\mathbf{d}}_{FDE-MF_ZF}^M = \mathbf{d} + \boldsymbol{\nu}_{FDE-MF-ZF}^M, \tag{2.42}$$

where $\boldsymbol{\nu}_{FDE-MF-ZF}^M = \mathbf{D}_{bM}^{-1}(\mathbf{AF}_{bM})^H \mathbf{W}_{NM} \boldsymbol{\Psi}^{-1} \mathbf{W}_{NM}^H \boldsymbol{\nu}$ is the enhanced noise vector. Since $\mathbf{D}_{bM}$ needs to be computed to obtain $\hat{\mathbf{d}}_{FDE-MF_ZF}^M$, it can be computed as,

$$\mathbf{D}_{bM} = \mathbf{F}_{bM} \mathbf{A}^H \mathbf{A} \mathbf{F}_{bM}^H. \tag{2.43}$$

Post-processing SNR for the lth symbol can be obtained as,

$$\gamma_{FDE-MF-ZF,l}^M = \frac{\sigma_{d^2}}{E[\boldsymbol{\nu}_{FDE-MF-ZF}^M (\boldsymbol{\nu}_{FDE-MF-ZF}^M)^H]_{l,l}}, \tag{2.44}$$

where the denominator in the above equation is the enhanced noise power for the lth symbol. In summary, BIDFT precoding can be understood by Figure 2.17(a).

2.7.2 DFT Precoded GFDM

DFT precoding has been used in OFDM systems [STB11, HT10]. It has been shown that DFT precoding reduces PAPR significantly and is one of the optimum precoding matrices to reduce PAPR in OFDM systems [Fal11]. This motivates us to investigate DFT precoded GFDM for reducing PAPR in GFDM systems. Suppose Q is spreading factor of the system, then DFT order/ size can be computed as $N_{DFT} = \frac{N}{Q}$. We assume that Q divides N completely. Two subcarrier mapping schemes are considered in this work [MLGnd] (i) localized frequency division multiple access (LFDMA) and (ii) interleaved frequency division multiple access (IFDMA).

Precoding matrix $\mathbf{P}$ can be defined as $\mathbf{P} = \mathbf{P}_m \mathbf{P}_c$, where $\mathbf{P}_c$ is a block diagonal matrix with each block being a DFT spreading matrix and $\mathbf{P}_m$ is a permutation matrix which implements subcarrier mapping, i.e., LFDMA or

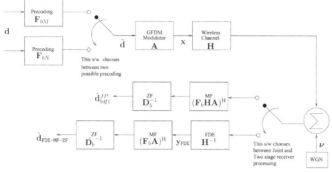

(a) BIDFT Precoded GFDM.

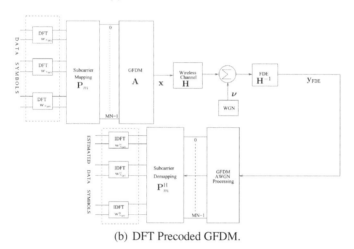

(b) DFT Precoded GFDM.

Figure 2.17   Block diagram of BIDFT and DFT precoded GFDM system.

IFDMA. The precoding spreading matrix $\mathbf{P}_c$ can be written as,

$$\mathbf{P}_c = \begin{bmatrix} \mathbf{W}_{N_{DFT}} & & & \\ & \mathbf{W}_{N_{DFT}} & & \\ & & \ddots & \\ & & & \mathbf{W}_{N_{DFT}} \end{bmatrix}, \qquad (2.45)$$

where $\mathbf{W}_{N_{DFT}}$ is the normalized DFT matrix of size $N_{DFT} \times N_{DFT}$. Permutation matrix $\mathbf{P}_m$ for LFDMA is an identity matrix. The DFT precoded GFDM system can be understood from Figure 2.17(b). The precoded data vector is GFDM modulated using the modulation matrix $\mathbf{A}$. The received signal can be equalized using a conventional linear [MKLFda, MMG$^+$14] or non-linear [DMLFda] equalizer. We will present here a ZF receiver for DFT precoded GFDM. The ZF equalized precoded data vector can be obtained as,

$$\hat{\mathbf{d}}_{zf} = (\mathbf{HA})^{-1}\mathbf{y} \;\; = \tilde{\mathbf{d}} + (\mathbf{HA})^{-1}\boldsymbol{\nu}. \tag{2.46}$$

The equalized data vector $\hat{\mathbf{d}}_{dft-spread}$ can be obtained as,

$$\hat{\mathbf{d}}_{dft-spread} = \mathbf{P}^{\mathrm{H}}\hat{\mathbf{d}}_{zf} \qquad = \underbrace{\mathbf{P}_c^{\mathrm{H}}}_{De-spreading} \times \underbrace{\mathbf{P}_m^{\mathrm{H}}}_{Subcarrier\,De-mapping} \times \hat{\mathbf{d}}_{zf}$$

$$= \mathbf{d} + \boldsymbol{\nu}_{dft-zf}, \tag{2.47}$$

where $\boldsymbol{\nu}_{dft-zf} = \mathbf{P}^{\mathrm{H}}((\mathbf{HA}))^{-1}\boldsymbol{\nu}$ is the post-processing noise vector. Post-processing SNR for the lth symbol can be obtained as,

$$\gamma_{dft-ZF,l}^{N} = \frac{\sigma_{d^2}}{E[\boldsymbol{\nu}_{dft-zf}^{N}(\boldsymbol{\nu}_{dft-zf}^{N})^{\mathrm{H}}]_{l,l}}, \tag{2.48}$$

where the denominator in the above equation is the enhanced noise power for the lth symbol.

2.7.3 SVD Precoded GFDM

The product of the channel matrix and the modulation matrix $(\mathbf{HA})$ can be decomposed as,

$$\mathbf{HA} = \mathbf{USV}^{H}, \tag{2.49}$$

where $\mathbf{U}$ and $\mathbf{V}$ are unitary matrices and $\mathbf{S}$ is a diagonal singular-value matrix, i.e., $\mathbf{S} = diag\{s_0, s_1 \cdots s_r, \cdots s_{MN-1}\}$, where s_r is the rth singular value. Then, $\mathbf{y}$ in (2.26) can be written as,

$$\mathbf{y} = \mathbf{USV}^{H}\mathbf{Pd} + \boldsymbol{\nu}. \tag{2.50}$$

At the transmitter, by choosing $\mathbf{P} = \mathbf{V}$ (assuming ideal feedback channel) and multiplying both sides with $\mathbf{U}^{H}$, the estimated symbol vector can be written as,

$$\hat{\mathbf{d}}^{svd} = \mathbf{Sd} + \mathbf{U}^{H}\boldsymbol{\nu}. \tag{2.51}$$

The estimated l^{th} symbol is then given by,

$$\hat{d}_l^{svd} = s_l d_l + \sum_{q=0}^{MN-1} [\mathbf{U}^{H}]_{l,q}\nu(q). \tag{2.52}$$

From the above, it can be seen that by using SVD-based precoding, interference is completely removed by orthogonalization of $\mathbf{HA}$ without the need for matrix inversion, which is required in ZF and minimum mean-square-error (MMSE) receiver. SINR for the lth symbol, can be computed as,

$$\gamma_l^{svd} = \frac{\sigma_d^2}{\sigma_\nu^2}|s_l|^2. \tag{2.53}$$

2.8 FBMC

The idea of FBMC dates back to 1966 when Chang conceptualized cosine modulated tone (CMT) as a method to transmit PAM modulated data over band-limited multiple subcarriers at Nyquist rate [Cha66]. A year later, in 1967, Stalzberg proposed a different version of FBMC which transmits QAM modulated data over band-limited multiple subcarriers at Nyquist rate [Sal67b], known as staggered modulated tone (SMT). The subcarrier distance is maintained at $\frac{1}{2T}$ (where T is the symbol duration), in CMT, to maintain Nyquist data rate and vestigial sideband modulation (VSB) is used to enforce orthogonality among subcarriers. Since the complex data symbol rides each subcarrier in SMT, the subcarrier distance of $\frac{1}{T}$ is used. This means that each subcarrier is DSB-SC modulated. To maintain orthogonality, each complex symbol for a subcarrier is offset QAM modulated as well as adjacent subcarriers are staggered oppositely. However, recently, in 2010, Behrouz Farhang has shown that these two schemes are only slightly different and can be obtained via each other by a multiplication of a constant in the frequency domain [FBGY10]. This also means that both CMT and SMT have similar performances in real transmission channels, which is further emphasized in [FBGY10].

SMT can be implemented using polyphase structure of filter banks [Hir81]. The polyphase approach has been the most used for FBMC as it provides a linear block approach. Using this structure, many receiver designs have been implemented, the most significant of which are by [ISRR07] and [ISR05]. However, the most efficient implementation of the FBMC based on frequency domain sampling was proposed by Bellanger in [MTB12]. This implementation provides the simplest form of FBMC receiver in the frequency domain. Its performance has been analyzed in [MTB13] and [MTB15a]. Also, the effects of timing and frequency offsets have been studied in [LGS14] and equalizers for combating them have been designed in [MTB15b].

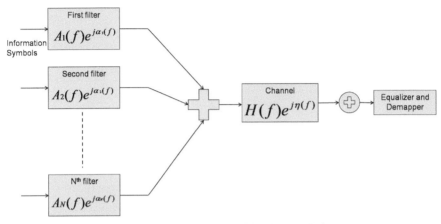

Figure 2.18 Block diagram of CMT transmission.

2.8.1 Cosine Modulated Tone

In this section, we will derive the continuous time domain CMT. Suppose we have N parallel AM channels as depicted in Figure 2.18, which can transmit through a band-limited transmission medium at a maximum possible data rate without ISI and ICI. The transmission filters $g_l(t)$ have gradual cut-off amplitude characteristics. The data rate per subcarrier is $\frac{1}{T}$, where T is the symbol duration. Hence, the subcarrier bandwidth is considered as $f_s = \frac{1}{2T}$. The center frequency for the lth subcarrier can be given as, $f_l = \frac{f_s}{2} + l f_s$. In the original work, Chang has considered a linear transmission medium but later works on CMT have consider an ideal channel. So, to be in sync with existing literature, we are also considering an ideal channel. Let the frequency response of the lth transmission filter be $G_l(f)e^{j\alpha_l(f)}$, where $G_l(f)$ and $\alpha_l(f)$ are the amplitude and phase response of the lth transmission filter, where $l = 0, 1, \ldots, N - 1$. Let $d_{m,l}$ be a real-valued data symbol for the $(m, l)^{\text{th}}$ time-frequency slot, where $m = -\infty, \ldots, -1, 0, 1, \ldots, \infty$ and $l = 0, 1, \ldots, N - 1$. Subcarriers are assumed to overlap into adjacent subcarriers as depicted in Figure 2.19.

Let us first concentrate on a single subcarrier. For the lth subcarrier, the received signal at the output of transmission medium can be given as,

$$y_l(t) = \sum_{m=-\infty}^{\infty} d_{m,l}g_l(t - mT) \tag{2.54}$$

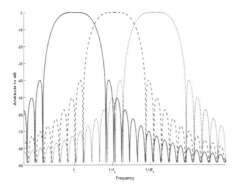

Figure 2.19 Overlapping subcarriers in CMT.

Now the received signal overlaps in time; they will be orthogonal if the signals corresponding to different data symbols are orthogonal, i.e.,

$$\int_{-\infty}^{\infty} g_l(t)g_l(t-mT)dt = 0 \ \text{ for } \ m = -\infty, \dots, \ -1, \ 1, \dots, \infty \ \text{ and}$$

$$k = 0, \dots, N-1. \tag{2.55}$$

This is the condition for ISI-free communication. Now let us consider a different r^{th} subcarrier. The received signal can be given as,

$$y_r(t) = \sum_{m=-\infty}^{\infty} d_{m,r}g_r(t-mT) \tag{2.56}$$

Now, this signal from the rth subcarrier will overlap with signal from the lth subcarrier. These signals will be orthogonal if

$$\int_{-\infty}^{\infty} g_l(t)g_r(t-mT)dt = 0 \ \text{ for } \ m = -\infty, \dots, \ -1, \ 0, \ 1, \dots, \infty \ \text{ and}$$

$$l, r = 0, \dots, N-1, \ l \neq r. \tag{2.57}$$

Using Fourier domain representation of $g_l(t)$, (2.55) can be rewritten as,

$$\int_{-\infty}^{\infty} G_l^2(f)e^{-j2\pi fmT}df = 0 \ \text{ for } \ m = -\infty, \dots, \ -1, \ 1, \dots, \infty \ \text{ and}$$

$$l = 0, \dots, N-1. \tag{2.58}$$

Let C be a real and positive constant and if $\int_{-\infty}^{\infty} C \cos(2\pi f mT)df = 0$, then this implies that $\int_{-\infty}^{\infty} C \sin(2\pi f mT)df = 0$. Since $G_l(f)$ is real and positive, using this theorem the above condition for ISI-free communication leads to

$$\int_{-\infty}^{\infty} G_l^2(f) \cos(2\pi f mT)df = 0 \text{ for } m = -\infty, \ldots, -1, 1, \ldots, \infty \text{ and}$$

$$l = 0, \ldots, N - 1.$$

(2.59)

In the above equation, since cosine is an even function around 0, values for positive k's will be the same for negative k's. Interestingly $G_l(f)$ is also an even function, the above equation can be simplified as,

$$\int_{0}^{\infty} G_l^2(f) \cos(2\pi f mT)df = 0 \text{ for } m = 1, 2 \ldots, \infty \text{ and } l = 0, \ldots, N-1.$$

(2.60)

In the same way, (2.57) can be written as,

$$\int_{-\infty}^{\infty} G_l(f)G_r(f)e^{-j(\alpha_l(f)-\alpha_r(f))2\pi f mT}df = 0$$

$$\text{for } m = -\infty, \ldots, -1, 0, 1, \ldots, \infty$$

$$\text{and } l, r = 0, \ldots, N - 1, l \neq r.$$

(2.61)

Now following the same steps as followed for the ISI-free condition, the ICI-free condition can be given as,

$$\int_{0}^{\infty} G_l(f)G_r(f) \cos(\alpha_l(f) - \alpha_r(f)) \cos(2\pi f mT)df = 0$$

(2.62)

$$\int_{0}^{\infty} G_l(f)G_r(f) \sin(\alpha_l(f) - \alpha_r(f)) \sin(2\pi f mT)df = 0$$

(2.63)

where $m = 0, 1, \ldots, \infty$ and $l, r = 0, \ldots, N - 1, l \neq r$.

2.8.2 Filter Characteristics

For ISI- and ICI-free communication, the transmitter filter $g_l(t)$, $\forall l$, should satisfy equations (2.60), (2.62), and (2.63). Let us assume that the transmitter filters are spread only to the adjacent subcarriers, i.e.,

$$G_l(f) = \begin{cases} \Gamma_l(f) & f_l - f_s \leq f \leq f_l + f_s \\ 0 & \text{otherwise} \end{cases}.$$

(2.64)

The ISI-free condition given in (2.60) can be written as,

$$\int_{f_l-f_s}^{f_l+f_s} \Gamma^2(f) \cos(\frac{\pi f m}{f_s}) df = 0 \text{ for } m = 1, 2 \dots, \infty \text{ and } l = 0, \dots, N-1.$$

(2.65)

The cosine term in the above equation, $\cos(\frac{\pi f m}{f_s})$, is even around $f_l - \frac{f_s}{2}$ as well as $f_l + \frac{f_s}{2}$. To make the equation zero, the function $\Gamma^2(f)$ should be odd around $f_l - \frac{f_s}{2}$ as well as $f_l + \frac{f_s}{2}$. But $\Gamma_l^2(f)$ is always greater than zero and hence it can be written as $\Gamma_l^2(f) = C_l + Q_l(f)$, where C_l and $Q_l(f)$ are a positive constant and a pure AC component for the lth subcarrier, respectively. Now, it can be very easily shown that $\int_{f_l-f_s}^{f_l+f_s} C_l \cos(\frac{\pi f m}{f_s}) df = 0$, $\forall m, l$. Hence, to satisfy (2.65), $Q_l(f)$ should be odd about $f_l - \frac{f_s}{2}$ and $f_l + \frac{f_s}{2}$, i.e.,

$$Q_l[(f_l + \frac{f_s}{2}) + f'] = -Q_l[(f_l + \frac{f_s}{2}) - f'], \ 0 < f' < \frac{f_s}{2} \qquad (2.66)$$

$$Q_l[(f_l - \frac{f_s}{2}) + f'] = -Q_l[(f_l - \frac{f_s}{2}) - f'], \ 0 < f' < \frac{f_s}{2} \qquad (2.67)$$

Next consider ICI, using the assumption $G_l(f) = 0$, for $f < f_l - f_s$, $f > f_l + f_s$, it can be shown that

$$G_l(f)G_r(f) = 0 \text{ for } r = l \pm 2, \ l \pm 3, \dots. \qquad (2.68)$$

This implies that ICI conditions given in (2.62) and (2.63) are satisfied for $r = l \pm 2, \ l \pm 3, \dots, \ \pm \infty$. It remains to investigate the ICI conditions for $r = l \pm 1$. Now, let us consider the case, $r = l + 1$. Equation (2.62) can be written as,

$$\int_0^\infty G_l(f)G_{l+1}(f) \cos(\alpha_l(f) - \alpha_{l+1}(f)) \cos(2\pi f m T) df \qquad (2.69)$$

$$= \int_{f_l}^{f_l+f_s} \sqrt{(C_l + Q_l(f))} \sqrt{(C_{l+1} + Q_{l+1}(f))} \cos(\alpha_l(f)$$

$$- \alpha_{l+1}(f)) \cos(\frac{\pi f m}{f_s}) df.$$

In the above equation, the term $\cos(\frac{\pi f m}{f_s})$ is even around $f_l + \frac{f_s}{2}$ and it can be very easily shown that $\sqrt{(C_l + Q_l(f))} \sqrt{(C_{l+1} + Q_{l+1}(f))}$ is an even function around $f_i + \frac{f_s}{2}$ (using (2.66) and (2.67)). It is required that $\cos(\alpha_l(f) - \alpha_{l+1}(f))$ is odd around $f_i + \frac{f_s}{2}$ for zero ICI condition. Let

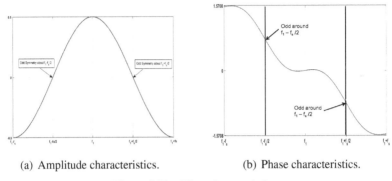

(a) Amplitude characteristics.   (b) Phase characteristics.

Figure 2.20   Filter characteristics.

$\alpha_l(f) - \alpha_{l+1}(f) = \theta + \gamma_l(f)$, where $\theta_l \in \{0, 2\pi\}$ is a constant angle and $\gamma_l(f)$ is an arbitrary function. Let $f_t \in \{-\frac{f_s}{2}, \frac{f_s}{2}\}$ be a frequency variable, it is needed that $\cos(\theta + \gamma_l(f_l + \frac{f_s}{2} + f_t)) = -\cos(\theta + \gamma_l(f_l + \frac{f_s}{2} - f_t))$. Since cosine function is odd around $\pm\frac{\pi}{2}$, take $\theta = \pm\frac{\pi}{2}$. It can be easily verified that $\cos(\alpha_l(f) - \alpha_{l+1}(f))$ will be odd around $f_l + \frac{f_s}{2}$ iff $\gamma_l(f)$ is odd around $f_l + \frac{f_s}{2}$. In a similar manner, for $r = l - 1$, it can be written as $l = r + 1$, and the same conditions can be obtained for the rth subcarrier. In conclusion, for zero ICI condition, the phase characteristics, $\alpha_l(f)$, should be shaped such that

$$\alpha_l(f) - \alpha_{l+1}(f) = \pm\frac{\pi}{2} + \gamma_l(f), \quad f_l < f < f_l + f_s, \quad l = 1, 2, \ldots, N - 1,$$
$$(2.70)$$

where $\gamma_l(f)$ is an arbitrary phase function with odd symmetry around $f_l + \frac{f_s}{2}$. Figure 2.20 illustrates the filter characteristics.

2.8.3 Simplified Filter Characteristics

In the previous section, filter characteristics, for communication over multiple band-limited subcarrier without ISI and ICI, are derived and can be followed using (2.66), (2.67), and (2.70). In this section, these conditions are simplified which are stated below.

1. C_l is the same for all l.
2. $Q_l(f)$ are identical pulse shaped, i.e.,

$$Q_{l+1}(f) = Q_l(f - f_s), \quad l = 0, 1, \ldots, N - 1. \qquad (2.71)$$

Using this and above assumption, it is straightforward to show that $G_l(f)$'s are identically pulse shaped, i.e.,

$$G_{l+1}(f) = G_l(f - f_s), \quad l = 0, 1, \ldots, N - 1. \tag{2.72}$$

The amplitude response of the lth filter can be given as,

$$G_l(f) = G_0(f - lf_s), \tag{2.73}$$

where $\alpha_0(f)$ is the amplitude response of the filter corresponding to the first subcarrier ($N = 0$) centered at $f_0 = \frac{1}{4T}$.

3. $Q_l(f)$ is even around f_l, i.e.,

$$Q_l(f_l + f') = Q_l(f_l - f'), \quad 0 < f' < f_s, \quad l = 0, 1, \ldots, N - 1. \tag{2.74}$$

This also means that $G_l(f)$ is also even around f_l, i.e.,

$$G_l(f_l + f') = G_l(f_l - f'), \quad 0 < f' < f_s, \quad l = 0, 1, \ldots, N - 1. \tag{2.75}$$

4. $\alpha_l(f)$ can be given as,

$$\alpha_l(f) = l\frac{\pi}{2} + \alpha_0(f - lf_s), \quad l = 0, 1, \ldots, N - 1, \tag{2.76}$$

where $\alpha_0(f)$ is the phase response of the filter corresponding to the first subcarrier ($N = 0$) centered at $f_0 = \frac{1}{4T}$. Since $g_l(t)$ is real, $\alpha_l(f)$ is odd around f_l. Now, zero ICI condition given in (2.70) is satisfied if and only if $\alpha_l(f)$ is odd around $f_l + \frac{f_s}{2}$ and $f_l - \frac{f_s}{2}$. A special case of this will be when $\alpha_0(f) = 0$ and subsequently $\alpha_l(f) = l\frac{\pi}{2}$ for $l = 0, 1, \ldots, N - 1$. This condition is fulfilled when $g_0(t)$ satisfies symmetric condition, i.e., $g_0(t) = g_0(-t)$.

Figure 2.21 shows the power spectral density of different filters which fulfills the above-mentioned filter characteristics. Phydyas [Bel01] filter has the lowest OoB radiation and generally preferred for FBMC transmission.

2.8.4 MMSE Equalizer for FBMC

The receiver design for FBMC is related to the MMSE criterion [WBN08, IL09]. For this design, we can write the transmitted signal of FBMC in an alternate form as,

$$s[n] = \sum_{k=0}^{M-1} y_k[n] \tag{2.77}$$

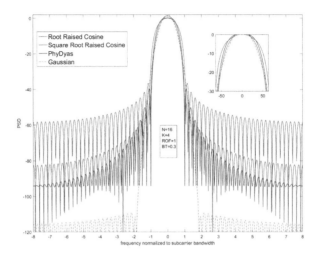

Figure 2.21   Prototype filters for CMT.

where the output for each subcarrier $y_k[n]$ is given as,

$$y_k[n] = \sum_{i=0}^{M-1} \sum_{l=-\inf}^{\inf} q_{ki}[l]v_i[n-l] + \eta_k[n] \qquad (2.78)$$

where $v_i[n]$ denotes the OQAM input symbols and $q_{ki}[n]$ is given as,

$$q_{ki}[n] = (g_i \star h \star f_k)_{\downarrow \frac{M}{2}} \qquad (2.79)$$

Here, $g_i[n]$ and $f_k[n]$ are the transmitter and receiver filters for the corresponding subcarrier indices. This equation represents the total signal processing involved in one subcarrier chain and the output obtained can be equalized directly. The equalizer mentioned here is a linear filter whose coefficients are designed according to the MMSE criterion. At first, L_{eq} samples of the output are collected and the filter coefficients are then multiplied with it to obtain the output. The output can be represented as,

$$y_k[n] = \sum_{i=k-1}^{k+1} Q_{ki}v_i[n] + B_k\eta[nM/2] \qquad (2.80)$$

where Q_{ki} is given as,

$$
Q_{ki} = \begin{bmatrix}
q_{ki}[0] & q_{ki}[1] & \cdots & q_{ki}[L_q] & 0 & \cdots & 0 \\
0 & q_{ki}[0] & q_{ki}[1] & \cdots & q_{ki}[L_q] & \ddots & \vdots \\
\vdots & \ddots & \ddots & \ddots & \ddots & \ddots & 0 \\
0 & \cdots & 0 & q_{ki}[0] & q_{ki}[1] & \cdots & q_{ki}[L_q]
\end{bmatrix}
$$

which is the convolution matrix formed by elements of q_{ki}, represented as $[q_{ki}[0], q_{ki}[1], ..., q_{ki}[L_q]]$ where $L_q + 1$ is the length of the convolution filter. B_k is the convolution and downsampling matrix. For a particular subcarrier k, only the contribution of the adjacent subcarriers $k - 1$ and $k + 1$ is taken into account. So, there are only three terms in the summation instead of M. $y_k[n]$ can be written as $y_k = [y_k[n], y_k[n-1], ...y_k[n - L_{eq} + 1]]$ which consists of Leq samples. The input samples contributing to the output are $v_k[n] = [v_k[n], v_k[n-1], ...,v_k[n - L_q - L_{eq} + 1]]$. This indicates that each sample at the output has a contribution from L_q input samples.

Equation (2.80) can also be written considering $v_i[n]$ as purely real symbols. This way, the purely imaginary symbols are converted into real and the $j = \sqrt{-1}$ factor is multiplied with the corresponding column of Q_{ki}. This results in,

$$
y_k[n] = \sum_{i=k-1}^{k+1} \overline{Q}_{ki} \overline{v}_i[n] + B_k \eta[nM/2] \tag{2.81}
$$

where $\overline{Q}_{ki}$ is the modified Q_{ki} with alternate purely real and purely imaginary columns and $\overline{v}_i[n]$ is purely real. If the equalizer is denoted by w_k, then the output after equalization is,

$$
Re(\tilde{v}_k[n]) = Re(w_k^H y_k[n]) \qquad for \ k + n \ even \tag{2.82}
$$

$$
Im(\tilde{v}_k[n]) = Im(w_k^H y_k[n]) \qquad for \ k + n \ odd \tag{2.83}
$$

The equalizer is designed according to the MMSE criterion which is given as,

$$
\min_{w_k} E[|e_k|^2] = \min_{w_k} E[|\overline{w}_k^T \overline{y}_k[n] - v_k[n - \delta]|^2] \tag{2.84}
$$

where $\overline{w}_k[n] = \begin{bmatrix} Re(w_k[n]) \\ Im(w_k[n]) \end{bmatrix}$ and $\overline{y}_k[n] = \begin{bmatrix} Re(y_k[n]) \\ Im(y_k[n]) \end{bmatrix}$. The MMSE equalizer can be defined as,

$$
\overline{w}_k^T = 1_{k,\delta}^T \overline{\overline{Q}}_{kk}^T \left(\sum_{i=k-1}^{k+1} \overline{\overline{Q}}_{ki} \overline{\overline{Q}}_{ki}^T + \frac{1}{SNR} \overline{B}_k \overline{B}_k^T \right)^{-1} \tag{2.85}
$$

Here, δ is the equalizer delay and it depends on the transmitter and receiver filter lengths and the channel length. $\overline{\overline{Q}}_{ki}$ is given as, $\overline{\overline{Q}}_{ki} = \begin{bmatrix} Re(\overline{Q_{ki}}) \\ Im(\overline{Q_{ki}}) \end{bmatrix}$ and $\overline{B_k} = \begin{bmatrix} Re(B_k) & -Im(B_k) \\ Im(B_k) & Re(B_k) \end{bmatrix}$.

2.9 UFMC

In UFMC, instead of pulse shaping every subcarrier individually, a group of subcarriers are pulse shaped together to form a resource block (RB) in the frequency domain. Each RB has a nearly flat frequency response within it and can be obtained by simple IDFT operation and filtering. There is some overlap between the RBs but it is negligible due to high sidelobe attenuation. So, UFMC is a near perfect orthogonal system when operated under normal conditions. Since the pulse shape is wider than FBMC, the time domain response is shorter in length. Hence, UFMC spectral efficiency is higher. Also the symbols do not overlap in time and block-wise decoding is possible.

UFMC was proposed as a new multicarrier technique for CoMP systems in [VWS+13b]. There it was shown to have better performance than OFDM with respect to SER in a multiple user scenario without and with the effect of CFO. It was also shown that UFMC has a better sidelobe performance than OFDM. UFMC has been developed using Dolph–Chebyshev filter which offers high sidelobe attenuation for small filter lengths because the sidelobe level can be controlled manually [Lyn97]. Hence, the length of this filter, the sidelobe level, and the bandwidth are user controllable parameters and one of them can be obtained using the other two. This makes the design of UFMC highly flexible. Authors in [WWSFdS14] have analyzed the effect of CFO on UFMC and have derived a new optimized filter design to maximize and minimize the inter-RB interference under CFO. This has been done by maximizing the signal to inter-RB interference ratio. In [WWS15], a similar approach is taken to optimize the filter for time-frequency errors jointly. It has been observed that the redesigned filter has reduced the main lobe width at the cost of high sidelobes, thus offering good immunity against inter-block interference in case of timing and frequency offsets. UFMC has been specially investigated for its adoption in CR spectrum due to its excellent sidelobe attenuation. A cancellation carrier-based interference cancellation scheme has been developed in [WZZW15], where protection to a primary user is given with the help of cancellation pilots at both ends of the occupied bandwidth. Channel estimation for UFMC systems have also been considered

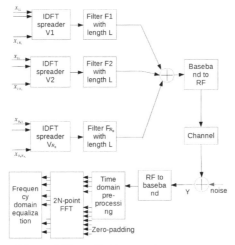

Figure 2.22   UFMC transceiver structure.

in [WWStB15] where it has been shown that the scheme is identical to OFDM under certain assumptions.

2.9.1 Structure of UFMC Transceiver

The UFMC transceiver is shown in Figure 2.22. Let R_B be the number of RBs with each having N_i subcarriers as input with i varying from 1 to R_B. The total number of subcarriers is $N = \sum_{i=1}^{R_B} N_i$. The filtering can be done independently for each RB, and hence the filter length depends on the width of the RB which in turn depends on the number of subcarriers. The filter lengths should be such that

$$N_1 + L_1 - 1 = N_2 + L_2 - 1 = N_3 + L_3 + 1........ = N_{R_B} + L_{R_B} - 1$$

for all the RBs. In the transmitter, at first an N-point IDFT(N is the number of subcarriers) is taken for the subcarriers in each RB. The input has N_i samples and the output has N samples. The IDFT matrix is a partial matrix of the full IDFT matrix with the columns corresponding to the indices of the subcarriers in the ith RB. After IDFT, the samples are convolved with the corresponding RB filter of length L_i. Hence, the total number of samples at output is $N + L_i - 1$. Here, the filter length is the same for all blocks, and hence the output signal length is of length $N + L - 1$. The filter is designed in such a way that there is a rapid fall in time domain in the first $\frac{L}{2}$ and last $\frac{L}{2}$ samples. This means that these samples can be considered to have negligible values and this serves as a padding between successive symbols. In other words, the time domain behavior of UFMC is almost similar to OFDM, where in place of OFDM's cyclic prefix, we have zero padding in UFMC. So, as long as the channel length is less than the filter length, there will be negligible or no ISI for both cases.

The filter used is an odd-order Dolph–Chebyshev filter which is designed according to the number of subcarriers in the RB [Lyn97]. The filter takes three arguments, length, passband bandwidth, and stopband attenuation which then can be designed according to the requirements and provides the filter impulse response. This gives a high level of control over the sidelobe attenuation for UFMC without having a large symbol length.

In the receiver, the received signal is first windowed using an appropriate filter (most preferably raised cosine filter) to contain it within a particular bandwidth. It is then zero-padded to make the total length $2N$. Then a $2N$-point DFT is taken. In the frequency domain, the $2N$ samples are alternatively selected, i.e., only the even indexed samples are selected. These samples carry the original data multiplied with the frequency domain channel coefficient. The odd-indexed samples carry the interference due to imperfect size of the DFT ($2N$ point DFT taken over $N + L - 1$ length symbol). The channel is then equalized in the frequency domain.

2.9.2 System Model for UFMC

Here, the UFMC transceiver is expressed as a linear model. A simpler form of a linear model is given in [SW14] but here we consider the detailed structure. The received signal after passing through the channel is given as,

$$\mathbf{Y} = \mathbf{H}_{(N+L+L_{ch}-2)\times(N+L-1)}\mathbf{\Phi}_{(N+L-1)\times((N+L-1)R_B)}$$
$$\mathbf{F}_{((N+L-1)R_B)\times NR_B}\mathbf{V}_{NR_B\times N}\mathbf{X}_{N\times 1} + \eta \qquad (2.86)$$

where $\mathbf{X}$ is the input data and is given as,

$$\mathbf{X} = [X_{11}, X_{12}, ..., X_{1K_1}, X_{21}, X_{22}, ..., X_{2K_2}, ...X_{R_B1}, ..., X_{R_BK_{R_B}}]^T \qquad (2.87)$$

Here, $K_1, K_2, ...K_{R_B}$ are the number of subcarriers in the R_B RBs. In this implementation, all RBs have the same number of subcarriers.

$\mathbf{V}$ is the combined IDFT matrix which is a block diagonal matrix given as $diag(\mathbf{V_1}, \mathbf{V_2}..........\mathbf{V_{R_B}})$ where each $\mathbf{V_i}$ is an IDFT matrix of size $N \times K_i$ where only those columns of an $N \times N$ IDFT matrix are taken which correspond to the subcarrier indices of the RB under consideration. V_i

is represented as,

$$\mathbf{V_i} = \begin{bmatrix} 1 & 1 & 1 & \cdots & 1 \\ e^{\frac{j2\pi i_1}{N}} & e^{\frac{j2\pi i_2}{N}} & e^{\frac{j2\pi i_3}{N}} & \cdots & e^{\frac{j2\pi i_{K_i}}{N}} \\ \vdots & \vdots & \vdots & \vdots & \vdots \\ e^{\frac{j2\pi(N-1)i_1}{N}} & e^{\frac{j2\pi(N-1)i_2}{N}} & e^{\frac{j2\pi(N-1)i_3}{N}} & \cdots & e^{\frac{j2\pi(N-1)i_{K_i}}{N}} \end{bmatrix} \qquad (2.88)$$

where $i_1, i_2 \ldots \ldots i_{K_i}$ correspond to the indices in the particular i^{th} RB.

$\mathbf{F}$ denotes the filtering matrix which is of $((N + L - 1)R_B) \times NR_B$. It is also a block diagonal matrix where each subblock is a toeplitz matrix denoting convolution. The filter length L is taken as the same for all RBs here. The sub-block $\mathbf{F_i}$ is given as,

$$\mathbf{F_i} = \begin{bmatrix} f_0 & 0 & 0 & 0 & \cdots & 0 & 0 \\ f_1 & f_0 & 0 & 0 & \cdots & 0 & 0 \\ \vdots & \vdots & \vdots & \vdots & \ddots & \vdots & \vdots \\ f_{L-1} & f_{L-2} & \cdots & f_0 & 0 & \cdots & 0 \\ \vdots & \vdots & \vdots & \vdots & \vdots & \vdots & \vdots \\ 0 & 0 & \cdots & f_{L-1} & \cdots & f_0 & 0 \\ \vdots & \vdots & \vdots & \vdots & \ddots & \vdots & \vdots \\ 0 & \cdots & \cdots & \cdots & \cdots & 0 & f_{L-1} \end{bmatrix} \qquad (2.89)$$

Here, $f_0, f_1, \ldots f_{L-1}$ are the filter coefficients. The filter coefficients are obtained by shifting the prototype filter to the center carriers of each RB. Thus, the filter coefficients for i^{th} resource block can be given as,

$$f_i = [a_0 \ a_1 \ \cdots a_{L-1}] \times \exp[j\frac{(i-1)K_i + (K_i/2)}{N}] \qquad (2.90)$$

Here, $[a_0, a_1, a_2 \ldots \ldots a_{L-1}]$ are the coefficients of the prototype filter. Φ is a concatenation of identity matrices which is given as,

$$\Phi = [\mathbf{I}_{(N+L-1)\times(N+L-1)} \quad \mathbf{I}_{(N+L-1)\times(N+L-1)} \\ \ldots \ldots \ldots \mathbf{I}_{(N+L-1)\times(N+L-1)}]. \qquad (2.91)$$

So, it consists of R_B identity matrices of size $N + L - 1$. Its purpose is to add the outputs of the RBs together.

$\mathbf{H}$ is the channel matrix of size $(N + L + L_{ch} - 2) \times (N + L - 1)$ and it is also a Toeplitz matrix. It is given as,

$$
\mathbf{H} =
\begin{bmatrix}
h_0 & 0 & 0 & 0 & \cdots & 0 & 0 \\
h_1 & h_0 & 0 & 0 & \cdots & 0 & 0 \\
\vdots & \vdots & \vdots & \vdots & \ddots & \vdots & \vdots \\
h_{L_{ch}-1} & h_{L_{ch}-2} & \cdots & h_0 & 0 & \cdots & 0 \\
\vdots & \vdots & \vdots & \vdots & \vdots & \vdots & \vdots \\
0 & 0 & \cdots & h_{L_{ch}-1} & \cdots & h_0 & 0 \\
\vdots & \vdots & \vdots & \vdots & \ddots & \vdots & \vdots \\
0 & \cdots & \cdots & \cdots & \cdots & 0 & h_{L_{ch}-1}
\end{bmatrix}
\tag{2.92}
$$

η is the additive white Gaussian noise vector.

Thus, the whole transmitter can be formulated as a linear model with the input as the data in the subcarriers and output in the time domain.

2.9.3 Output of the Receiver for the UFMC Transceiver Block Diagram

An analysis has been provided for the decoded symbols in the frequency domain at the receiver. By the decoding technique through $2N$-point FFT, the effective ICI from the subcarriers within the corresponding RB and all the other RBs at the even subcarriers at the output of the $2N$-point FFT becomes zero.

The received signal without noise is given as,

$$
\mathbf{Y}_{1 \times (N+L-1)} = \sum_{i=1}^{B} (\mathbf{f_i} * \mathbf{x_i})
\tag{2.93}
$$

where B is the number of RBs and f_i is the time domain filter for the ith RB. x_i is the IDFT of the data in the ith RB and is given as,

$$
x_i(l) = \frac{1}{N} \sum_{k \in S_i} X_i(k) exp(\frac{j2\pi lk}{N})
\tag{2.94}
$$

where S_i is the set of subcarrier indices belonging to the ith RB. If we consider the output of only the ith RB, then the mth element of the output is,

$$y_i(m) = \sum_{g=0}^{L-1} f_i(g)x_i(m-g) \tag{2.95}$$

At the output, we take a $2N$-point DFT and thus the k'^{th} element is given as,

$$Y_i(k') = \sum_{m=0}^{2N-1} y_i(m)e^{\frac{-j2\pi mk'}{2N}} \tag{2.96}$$

$$Y_i(k') = \sum_{m=0}^{2N-1}\sum_{g=0}^{L-1} f_i(g)x_i(m-g)e^{\frac{-j2\pi mk'}{2N}} \tag{2.97}$$

$$Y_i(k') = \frac{1}{N}\sum_{m=0}^{2N-1}\sum_{k\in S_i}\sum_{g=0}^{L-1} f_i(g)X_i(k)e^{\frac{j2\pi(m-g)k}{N}}e^{\frac{-j2\pi mk'}{2N}} \tag{2.98}$$

$$Y_i(k') = \frac{1}{N}\sum_{m=0}^{2N-1}\sum_{k\in S_i} X_i(k)F_i(k)e^{\frac{j2\pi mk}{N}}e^{\frac{-j2\pi mk'}{2N}} \tag{2.99}$$

Now if $k' = 2p$ or k' is even for $p \in S_i$, then we have,

$$Y_i(2p) = \frac{1}{N}\sum_{m=0}^{2N-1}\sum_{k\in S_i} X_i(k)F_i(k)e^{\frac{j2\pi m(k-p)}{N}} \tag{2.100}$$

So, if $k = p$, then $Y_i(2p) = 2X_i(p)F_i(p)$ and for all other k, $Y_i(2p)$ is zero. Hence, the subcarrier data multiplied with the filter are recovered at twice the transmitter subcarrier index.

For $k' = 2p + 1$ or k' is odd for $p \in S_i$, then,

$$Y_i(2p+1) = \frac{1}{N}\sum_{m=0}^{2N-1}\sum_{k\in S_i} X_i(k)F_i(k)e^{\frac{j2\pi mk}{N}}e^{\frac{-j2\pi m(2p+1)}{2N}} \tag{2.101}$$

In this equation, for any k, the data cannot be recovered in pure form and there is always interference from the other subcarriers in the block.

For the other RBs $j \neq i$, the particular output index for RB i will never match to any indices of the other RBs. Hence, the output of (2.99) will always be zero for even subcarriers and interference terms for odd subcarriers.

2.10 Performance Comparison

In this section, we present the performance comparison of the contending waveforms. We have done Monte Carlo simulations to compare the performance of waveforms. For evaluation, we have considered the channel model, given in [MMG$^+$14]. We have considered uncoded system and use of linear receiver structures. The parameters of considered waveforms are provided in Table 2.3. It is assumed that the subcarrier bandwidth is comparable to the coherence bandwidth of the channel for FSFC. SNR loss due to CP is also considered for FSFC.

Figure 2.23(a) shows the comparison of BER vs $\frac{E_b}{N_0}$ of considered waveforms for 16 QAM modulation. It is seen from the figure that BIDFT-GFDM outperforms other waveforms. This happens because in the case of BIDFT-GFDM waveform, the signal is spread in the entire operational bandwidth and hence exploits frequency diversity. SCFDE waveform which also spreads the signal in the entire bandwidth performs worse than BIDFT-GFDM because GFDM systems use lesser CP length than OFDM system and hence have spectral efficiency gains. This gain manifests better BER performance of GFDM-based systems than OFDM-based systems. IFDMA-GFDM performs

Table 2.3   Simulation parameters of comparison of waveforms

Parameters	Attributes
Number of subcarriers N	64
Number of time slots M (for GFDM)	5
Mapping	16 QAM
Pulse shape (GFDM)	RRC with ROF =0.1
Pulse shape (FBMC)	Phy-Dyas filter [VIS$^+$09]
Pulse shape (UFMC)	Equi-Ripple 120 dB attenuation FIR
Number of resource blocks (UFMC)	4
Block-based pulse shape for OFDM, SCFDMA, and SCFDE	RRC with ROF = 0.35 and filter length = 17
CP length N_{CP}	16
Channel length N_{ch}	16
Power delay profile	$[10^{-\frac{\alpha}{5}}]^{\mathrm{T}}$, where $\alpha = 0, 1 \cdots N_{ch} - 1$
Subcarrier bandwidth	3.9 KHz
RMS delay spread	4.3 μ s
Coherence bandwidth	4.7 KHz

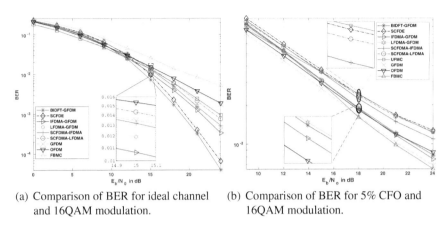

(a) Comparison of BER for ideal channel and 16QAM modulation.

(b) Comparison of BER for 5% CFO and 16QAM modulation.

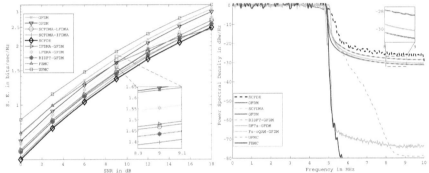

(c) Comparison of spectral efficiency for ideal channel using water filling algorithm.

(d) Comparison of OoB leakage.

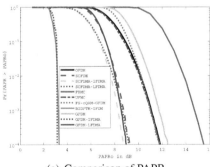

(e) Comparison of PAPR.

Figure 2.23 Comparison of waveforms.

better than LFDMA-GFDM because IFDMA-GFDM spreads the signal in a non-contiguous manner and LFDMA-GFDM spreads the signal locally and hence IFDMA-GFDM exploits more frequency diversity. IFDMA-SCFDMA performs better than LFDMA-SCFDMA due to the same reason.

UFMC performs better than vanilla forms of GFDM, FBMC, OFDM, and oQAM-GFDM. FBMC performs better than OFDM at low SNR values and worse at high SNR values.

Next, we will see the effect of CFO on the BER performance of considered waveforms. We have considered 5% CFO value, i.e., the frequency offset is 5% of subcarrier bandwidth. Figure 2.23(b) shows the comparison of BER versus $\frac{E_b}{N_0}$ of considered waveforms for 16 QAM modulation and 5% CFO. It is observed that FBMC outperforms all other waveforms in this case. GFDM performs better than precoded OFDM, precoded GFDM, OFDM, and UFMC. Precoded GFDM performs better than precoded OFDM. IFDMA-GFDM performs better than OFDM, which has a comparable performance to LFDMA-GFDM. LFDMA-GFDM performs better than BIDFT-GFDM. SCFDMA-IFDMA performs similar to UFMC and better than SCFDMA-LFDMA, which performs better than SCFDE. It is also observed that precoded GFDM waveforms and precoded OFDM waveforms show degradation of two order (100) in the case of CFO when compared with the ideal case (Figure 2.23(a)), whereas OFDM shows degradation one order (10) and GFDM and FBMC demonstrate the deterioration of lesser than one order. It can be inferred that the frequency spreading has the adverse effect on BER performance in the presence of CFO. BIDFT-GFDM as well SCFDE show the highest degradation as these waveforms are spread over the entire frequency range. Among DFT spreading waveforms, IFDMA-based waveforms degrade more than LFDMA waveforms as IFDMA-based waveforms spread the signal in a non-contiguous manner whereas LFDMA-based waveforms spread the signal in a contiguous manner. FBMC and GFDM are rather resilient toward CFO due to the use of very well localized pulse shapes in the frequency domain. UFMC, which filters a group of subcarriers, is less resistant toward CFO than FBMC and GFDM.

Figure 2.23(c) shows the spectral efficiency of considered waveforms using optimum per symbol power derived from water-filling algorithm [PS09]. We assume perfect channel knowledge both at the transmitter and at the receiver. It can be seen that UFMC has the highest spectral efficiency. GFDM has higher spectral efficiency than FBMC, OFDM, and precoded OFDM and precoded GFDM waveforms. FBMC has a higher spectral efficiency than OFDM and a comparable capacity to GFDM at low SNR.

At high SNR values, FBMC has a lower spectral efficiency than OFDM. OFDM has a higher spectral efficiency than precoded OFDM waveforms and precoded GFDM waveforms. LFDMA-based schemes are found to be better than IFDMA-based schemes for both GFDM and SCFDMA. SCFDE has a higher spectral efficiency than oQAM modulated GFDM waveforms. It can be concluded that SNR gain due to lesser CP length in GFDM manifests a higher spectral efficiency in GFDM-based waveforms than OFDM-based waveforms. It can also be concluded that waveforms that are exploiting frequency diversity provide lesser spectral efficiency than those who are not exploiting frequency diversity. FBMC and UFMC waveforms have spectral efficiency gains because they do not use CP. However, in the case of FBMC, subcarriers become non-orthogonal in FSFC, which degrades its SINR and hence spectral efficiency.

Figure 2.23(d) shows one-sided power spectral density plot of considered waveforms. In a 20-MHz system, 128 subcarriers are considered out of which 64 subcarriers are switched off (32 on each edge) to observe OoB characteristics of waveforms. Power spectral density is averaged over 10^4 transmitted symbols for each considered waveform. FBMC has the lowest stopband attenuation and narrowest transition band. UFMC has 58 and 63 dB more stopband attenuation than OFDM, respectively. UFMC has a very large transition band. GFDM has 5 dB more stopband attenuation and narrower transition band than OFDM. Precoded GFDM has quite similar stopband attenuation to GFDM and precoded OFDM also has quite similar stopband attenuation to OFDM, which shows that DFT-based precoding has a little effect on OoB characteristics in OFDM and GFDM. SCFDE has the worst stopband attenuation.

Figure 2.23(e) shows complementary cumulative distribution function plot of PAPR of all waveforms. 10^5 transmitted frames were generated, where each frame has four transmitted signal blocks. We compare the PAPR value that is exceeded with probability less than 0.01% ($\Pr\{PAPR > PAPRo = 10^{-4}\}$). It is observed that BIDFM-GFDM and IFDMA-GFDM have lowest PAPR that is lesser than OFDM by 8.5 dB. SCFDE, SCFDMA-IFDMA, GFDM-LFDMA, and SCFDMA-LFDMA have comparable PAPRs, which is smaller than OFDM by approximately 3.5 dB. Block-based pulse shapes used in SCFDE and SCFDMA reduce the PAPR gain in these systems [Sli02]. UFMC has a quite similar PAPR to OFDM. GFDM has 0.7 dB whereas FBMC, which has the highest PAPR, has 3 dB worse PAPR than OFDM. A review of the critical performance parameters for the different waveforms is given in Table 2.4.

Table 2.4 A comprehensive comparison of the different performance parameters of the waveforms considered

Parameter	OFDM	FBMC	GFDM	BIDFT-GFDM	LFDMA-GFDM	IFDMA-GFDM	SCFDMA-LFDMA	SCFDMA-IFDMA	SCFDE	UFMC
Out-of-band emission	Very high	Very low	High	High	High	High	Very high	Very high	Very high	Low
BER performance	Moderate	Poor	Moderate	Very good	Good	Good	Good	Good	Very good	Moderate
PAPR performance	Poor	Very poor	Poor	Very good	Good	Very good	Good	Good	Good	Poor
Resilience to carrier frequency offset	Poor	Very good	Very good	Very poor	Very poor	Very poor	Very poor	Very poor	Very poor	Good
Spectral efficiency	Good	Moderate	Good	Poor	Poor	Poor	Poor	Poor	Poor	Very good

Table 2.5 Waveform recommendations for different application requirements

Parameter	OFDM	FBMC	GFDM	BIDFT-GFDM	LFDMA-GFDM	IFDMA-GFDM	SCFDMA-LFDMA	SCFDMA-IFDMA	SCFDE	UFMC
High spectral efficiency	PR	PR	PR	NR	NR	NR	NR	NR	NR	HR
Good spectrum isolation	NR	HR	NR	NR	NR	NR	NR	NR	NR	PR
High reliability (low error)	NR	NR	NR	HR	PR	PR	PR	PR	HR	NR
Low latency applications	NR	NR	PR	PR	PR	PR	NR	NR	NR	HR
High CFO applications	NR	HR	HR	HR	NR	HR	NR	NR	NR	PR
Low-cost power amplifier	NR	NR	NR	HR	PR	PR	PR	PR	PR	NR

HR-highly recommended, PR-partially recommended, and NR-not recommended.

Based on the detail performance analysis of the contending waveforms, the following conclusions can be made.

In the case of the ideal channel, precoded GFDM has BER gain of more than two order (10^{-2}) better than plain GFDM, OFDM, UFMC, and FBMC. Regarding SNR, it is nearly 8 dB better than them. Single carrier-based schemes are relatively poorer than precoded GFDM because of CP loss. In the case of CFO, FBMC and plain GFDM are performing best. FBMC and plain GFDM achieve around 3 dB SNR gain over OFDM in the presence of CFO. UFMC provides the highest spectral efficiency that is followed by GFDM, OFDM, and lastly by FBMC at high SNR. FBMC has the lowest OoB leakage. PAPR of precoded GFDM is better than rest by several dB whereas FBMC has the highest PAPR.

Therefore, we can say that, for high spectral efficiency requirement, one may use UFMC, for high-reliability requirement, one may choose BIDFT-GFDM, for low PAPR requirement, one may choose BIDFT-GFDM or IFDMA-GFDM, for flexibility requirement, one may choose GFDM, for CFO resilient requirement, one may choose GFDM or FBMC, while for low OoB leakage requirement, one may choose FBMC. These conclusions can be followed by Table 2.5.

3

Non-orthogonal Multiple Access

Recently, it has been proposed that superposition coding (SC) with successive interference cancellation (SIC) could be a realization for non-orthogonal multiple access (NOMA) schemes for beyond 4G networks [Y. 13]. It is known that SC with SIC achieves the capacity of the degraded broadcast channel [CT06], i.e., it utilizes the available spectrum in the most efficient manner possible.

From information-theoretic point of view, orthogonal frequency division multiplexing-based NOMA (OFDM-NOMA) systems are equivalent to the parallel broadcast channel. The optimal resource allocation for the parallel broadcast channel is given by David Tse in his seminal work [Tse97]. The same resource allocation approach also applies to OFDM-NOMA systems. However, this scheme may lead to significant increase in signaling overhead and the users may suffer from SIC error propagation during the decoding process. Hence, the resource allocation scheme with a restricted number of users are required to be considered [SHSB13, SBKN13, HK13, LHK14]. Resource allocation with the objective of maximizing weighted sum rate of the system for a multi-carrier system with a restricted number of users per sub-band is presented in this chapter.

A multi-cellular OFDM-NOMA system is equivalent to parallel interfering broadcast channel. For the simplest case of interference channel, the optimal rate region is characterized by Han–Kobayashi region [Kob81]. However, no practical scheme exists that achieves the Han–Kobayashi region. Hence, for an interfering broadcast channel, characterizing and finding an optimal rate region from an information theoretic point of view are far more challenging than the parallel broadcast channel. Usually, the resource allocation is performed by solving an optimization problem with appropriate constraints. The performance evaluation of the downlink of a multi-cellular OFDM-NOMA system based on the coordinated scheduling concept is also described here.

59

Finally, a method to efficiently select multiple users and perform non-iterative power allocation is described.

3.1 OFDM-based Non-orthogonal Multiple Access

We considered a downlink OFDM-NOMA-based cellular system with a hexagonal grid layout of 19 cells. Each cell is divided into three sectors with directional antennas. M number of users are deployed uniformly in the center cell (known as desired cell) and rest of the cells contribute to the inter-cell interference (ICI). The total bandwidth of the system (B_{sys}) is equally partitioned into N number of sub-bands of bandwidth SB_{BW}. Moreover, each sub-band consists of N_{sc} number of subcarriers of bandwidth SC_{BW} in the frequency domain and spans over one transmission time interval (T_{tti}) in time domain. Since the system employs multiplexing of users through SC, a single sub-band can be assigned to multiple users.

Considering that M users are multiplexed on the nth sub-band (SB_n), a superposition coded symbol (as in (3.1)) is transmitted by the base station on each of the subcarriers in SB_n.

$$\mathrm{x_n} = \sum_{i=1}^{M} \sqrt{\beta_i^n P_n} x_i, \tag{3.1}$$

where x_i's are complex QAM symbols drawn from a constellation with zero mean and unit variance, P_n is the power allocated on SB_n, and β_i^n is the fraction of power allocated to i^{th} user multiplexed on SB_n such that $\sum_{i=1}^{M_r} \beta_i^n = 1$. If the ith user is not scheduled on SB_n then $\beta_i^n = 0$. The signal received by the mth user (UE_m) on SB_n is

$$y_m^n = h_m^n \mathrm{x_n} + v_m^n \tag{3.2}$$

$$= \sqrt{\beta_m^n P_n} h_m^n x_m + \sum_{i=1, i \neq m}^{M} \sqrt{\beta_i^n P_n} h_m^n x_i + v_m^n,$$

where h_m^n is the complex channel response between the desired base station and UE_m on SB_n. It captures the effect of path loss, shadowing, and small scale fading. v_m^n constitutes of both additive white Gaussian noise and ICI.

Without SIC, the received signal to interference plus noise ratio (SINR) (Pre-SIC SINR) of UE_m on SB_n

$$\text{SINR}_m^{n,Pre} = \frac{\beta_m^n P_n |h_m^n|^2}{\sum_{i=1, i \neq m}^{M} \beta_i^n P_n |h_m^n|^2 + E[|v_m^n|^2]}$$

$$= \frac{\beta_m^n P_n \Gamma_m^n}{1 + \sum_{i=1, i \neq m}^{M} \beta_i^n P_n \Gamma_m^n}, \tag{3.3}$$

where $\Gamma_m^n = \frac{|h_m^n|^2}{E[|v_m^n|^2]}$ represents the channel to interference plus noise ratio (CINR) of UE_m on SB_n. SIC thrives on the fact that if all the users having a poorer channel condition with respect to UE_m are able to decode their respective transmitted symbols, then UE_m would also be able to decode all the symbols transmitted to those users. Without loss of generality, consider that for all the users multiplexed over SB_n, the CINRs are ordered as $|\Gamma_1^n|^2 \geq |\Gamma_2^n|^2 \geq \dots \geq |\Gamma_m^n|^2 \geq |\Gamma_{m+1}^n|^2 \geq \dots \geq |\Gamma_M^n|^2$. Hence, UE_m would be able to successfully decode and remove the symbols of the users UE_{m+1}, UE_{m+2}, ..., UE_M. However, the interference power from UE_1, UE_2, ..., UE_{m-1}, who are having better channel conditions compared to UE_m, will be treated as noise by UE_m during the decoding process. After performing SIC, the post-processing SINR of UE_m on SB_n is given by

$$\text{SINR}_m^n(\boldsymbol{\beta}_n, P_n) = \frac{\beta_m^n P_n \Gamma_m^n}{1 + \sum_{i=1}^{m-1} \beta_i^n P_n \Gamma_m^n}, \tag{3.4}$$

where $\boldsymbol{\beta}_n = \begin{bmatrix} \beta_1^n & \beta_2^n & \cdots & \beta_M^n \end{bmatrix}^T$ contains the fraction of power allocated to each user.

Using (3.4), the total number of bits that can be transmitted to UE_m on SB_n during a time slot can be written as

$$R_m^n(\boldsymbol{\beta}_n, P_n) = \text{SB}_{\text{BW}} T_{\text{tti}} \log_2(1 + \text{SINR}_m^n(\boldsymbol{\beta}_n, P_n)). \tag{3.5}$$

The objectives of scheduling and resource allocation unit are to (i) select the best set of co-channel users for each sub-band, (ii) distribute power among the multiplexed users on each sub-band, and (iii) allocate power across the sub-bands subject to maximum power (P_{max}) constraint so that the total system utility (weighted sum rate) can be maximized.

The optimal solution to the above-mentioned problem is presented in [Tse99]. Using the proposed optimal algorithm, given the weights and channel condition of the users in the system, a set of rate vectors and corresponding power allocation among users and across sub-bands is obtained which maximizes the instantaneous weighted sum rate of the system. Although the optimal power allocation and user selection algorithm achieves the maximum gain for OFDM-NOMA systems, it may not be practical to implement in the real world due to the following reasons:

1. It has no limitations on the maximum number of users which can be multiplexed over a sub-band. Hence, the amount of required signaling overhead required may become too high.

2. Due to higher number of multiplexed users over a sub-band, there is a higher chance of error propagation due to incorrect SIC decoding, which would degrade the system performance.

Hence, in order to reduce the signaling overhead and chances of error propagation due to incorrect SIC decoding, a natural choice is to limit the number of multiplexed users over a sub-band to two or three. The choice of limiting the number of multiplexed users to two or three is well supported by the results presented in [ZSAA13] and [SBK13]. Therefore, the above-described problem considering a restricted number of users per sub-band to M_r can be written as,

$$\text{maximize} \quad \sum_{n=1}^{N} \sum_{i=1}^{M} \alpha_i^n w_i R_i^n (\beta_n, P_n), \quad (3.6a)$$

$$\text{subject to} \quad \sum_{n=1}^{N} P_n = P_{max},$$

$$P_n \geq 0, \quad (3.6b)$$

$$\|\beta_n\|_1 = 1, \quad (3.6c)$$

$$\sum_{i=1}^{M} \alpha_i^n \leq M_r, \quad (3.6d)$$

where the last constraint (3.6d) ensures that the maximum number of scheduled users on a sub-band should not exceed M_r. Furthermore, α_i^n is a binary variable defined as follows

$$\alpha_i^n = \begin{cases} 1 & \text{if UE}_i \text{ is assigned to SB}_n \\ 0 & \text{Otherwise} \end{cases} \quad (3.7)$$

Based on the value of α_i^n, β_i^n takes the following values

$$\beta_i^n \begin{cases} > 0 & \text{if } \alpha_i^n = 1 \\ = 0 & \text{if } \alpha_i^n = 0 \end{cases} \tag{3.8}$$

Due to the complexity of the above problem, which is exponential in terms of number of users and sub-bands in the system, we have attempted to solve the problem in two steps.

1. In the first step, for each sub-band,

   (a) Find the set of users to be multiplexed over a sub-band (discussed in Section 3.1.1).

   (b) Find the power ratio vector ($\bar{\beta}_n = [\bar{\beta}_1^n, \bar{\beta}_2^n, \ldots, \bar{\beta}_{M_r}^n]$) based on which multiplexed users will be assigned fractions of powers available to the total resource band (discussed in Section 3.1.1).

   In the aforementioned two steps, the values of α_i^n and β_i^n are obtained.

2. In the second step, power allocation is performed across the sub-bands such that the total weighted instantaneous sum rate of the system gets maximized subject to maximum power constraint. (discussed in Section 3.1.1).

Once the set of multiplexed users and corresponding $\bar{\beta}_n$ is obtained from the first step, the allocated power across sub-bands can be obtained by solving the following optimization problem

$$\text{maximize} \qquad \sum_{n=1}^{N} \sum_{i=1}^{M_r} w_i^n R_i^n (P_n, \bar{\beta}_n), \tag{3.9a}$$

$$\text{subject to} \qquad \sum_{n=1}^{N} P_n = P_{max},$$

$$P_n \geq 0, \tag{3.9b}$$

where w_i^n is the weight of the ith user multiplexed over SB_n. M_r represents the number of multiplexed users over SB_n. Without loss of generality, we assume that there is one to one correspondence among users' weights w_i^n's and achievable instantaneous throughputs R_i^n's over SB_n. Furthermore, w_i^n's and R_i^n's are ordered in a decreasing manner as per users' channel conditions

(CINRs), i.e., w_i^n and $R_i^n(P_n, \beta_n)$ for $i = 1$ represent the weight and instantaneous datarate of the user who experiences the best channel condition (CINR) and removes the interference due to rest of the multiplexed users through SIC. Similarly, for $i = M_r$, the user weight and instantaneous rate correspond to the user who is experiencing the worst channel condition and hence treat the interference from rest of the users as noise. The performance of OFDM-NOMA system is evaluated by restricting M_r to a maximum value of two or three.

3.1.1 Algorithms for User Multiplexing and Power Allocation

The performance of NOMA is principally guided by the selection of user set over a particular sub-band and allocation of the power to the multiplexed users on the sub-band.

A Brief Description of User Multiplexing Algorithm
User Selection Through Exhaustive Search
Selection of the best set of users to be multiplexed is a combinatorial optimization problem and is also coupled with the power allocation algorithm among the multiplexed users. For a given power distribution algorithm, one approach to select the best set is to search over all possible combinations of users and select the one which maximizes the weighted sum rate on the sub-band as mentioned in [SBKN13].

The overall complexity of the algorithm would be $O(2^M)$ [CSRL01].

Heuristic Algorithm for User Selection
In order to reduce the complexity of the exhaustive search, the user selection can be performed such that the first user to be multiplexed over the sub-band would be the one which has the maximum weighted sum rate (similar to proportional fair scheduling [JPP00]). The next user is selected out of $M-1$ remaining users subject to the condition that when the power is divided among the users on that sub-band, the new set of multiplexed users should improve the total weighted sum rate over that sub-band. This process goes on iteratively for the selection of rest of the users to be multiplexed over the sub-band. The complexity of the user selection algorithm for a system consisting of M number of users and N number of sub-bands would be $(1 + (M_r - 1)T_1)O(MN)$ [CSRL01].

Power allocation among multiplexed users

In literature, such as [SBKN13] and [BLS+13], the FTPA, prevalent for power control in LTE uplink, is used for OFDM-NOMA downlink systems. The method described below is the DC programming approach which is detailed in [PD14].

DC programming-based power allocation

Let M_r be the maximum number of users to be multiplexed over SB_n. In this work, we have evaluated the system performance for $M_r = 2$ and 3. The scheduler's objective is to select the set of users and find the fraction of power to be allocated to each multiplexed users so that the weighted sum rate on that sub-band gets maximized. For a given set of users, the scheduler needs to solve the following optimization problem on each sub-band

$$\text{maximize} \quad \sum_{i=1}^{M_r} w_i^n \log_2 \left(1 + \frac{\bar{\beta}_i^n P_n \Gamma_i^n}{1 + \sum_{j<i} \bar{\beta}_j^n P_n \Gamma_i^n} \right) \tag{3.10a}$$

$$\text{subject to} \quad \|\bar{\beta}_n\|_1 = 1 \tag{3.10b}$$

$$\bar{\beta}_i^n \geq 0., \tag{3.10c}$$

where $\bar{\beta}_i^n$ is the fraction of the total power allocated to the ith user on SB_n. The equivalent convex formulation of the above problem is given as

$$\underset{\bar{\beta}_i^n \geq 0}{\text{minimize}} \sum_{i=1}^{M_r} w_i^n \log_2 \left(1 + \frac{\bar{\beta}_i^n P_n \Gamma_i^n}{1 + \sum_{j<i} \bar{\beta}_j^n P_n \Gamma_i^n} \right), \tag{3.11a}$$

$$\text{subject to} \quad \|\bar{\beta}_n\|_1 = 1. \tag{3.11b}$$

The objective function of the above problem can be represented as the difference of two functions of β_n, i.e.,

$$\sum_{i=1}^{M_r} -w_i^n \log_2 \left(\frac{1 + \sum_{k=1}^{i} \bar{\beta}_k^n P_n \Gamma_i^n}{1 + \sum_{j<i} \bar{\beta}_j^n P_n \Gamma_i^n} \right) \tag{3.12a}$$

$$= \sum_{i=1}^{M_r} -w_i^n \log_2 \left(1 + \sum_{k=1}^{i} \bar{\beta}_k^n P_n \Gamma_i^n \right)$$

$$- \left(-\sum_{i=1}^{M_r} w_i^n \log_2 \left(1 + \sum_{j<i} \bar{\beta}_j^n P_n \Gamma_i^n \right) \right) \tag{3.12b}$$

$$= \mathbf{f}(\bar{\beta}_n) - \mathbf{g}(\bar{\beta}_n), \tag{3.12c}$$

where both $\mathbf{f}(\bar{\beta}_n)$ and $\mathbf{g}(\bar{\beta}_n)$ being the sum of negative log functions are convex functions of $\bar{\beta}_n$. The methodology to solve the above optimization problem is presented in Algorithm 3.1. For the special case of two users per sub-band, Problem (3.11) can be written as

$$\operatorname*{argmin}_{\beta_n \in [0\ 1]} - w_1^n \log_2\left(1 + \bar{\beta}_1^n P_n \Gamma_1^n\right) - \left(-w_2^n \log_2\left(1 + \bar{\beta}_1^n P_n \Gamma_2^n\right)\right)$$
$$- w_2^n \log_2\left(1 + P_n \Gamma_2^n\right). \tag{3.13}$$

The value of β_1^n is

$$\bar{\beta}_1^n = \begin{cases} 0, & \text{if first derivative is negative both at } 0 \text{ and } 1 \\ 1, & \text{if first derivative is positive both at } 0 \text{ and } 1 \\ \dfrac{w_2^n \Gamma_2^n - w_1^n \Gamma_1^n}{P_n\left(w_1^n - w_2^n\right)\Gamma_1^n \Gamma_2^n}, & \text{else.} \end{cases}$$
$$\tag{3.14}$$

The value of $\bar{\beta}_2^n$ is $1 - \bar{\beta}_1^n$.

Power allocation across sub-bands
Once the information regarding the users to be multiplexed over a sub-band and the fraction of powers that needs to be allocated is available at the scheduler, it will try to maximize the weighted sum rate of the total system by judiciously allocating power to each sub-band subject to maximum power constraint. In the following two subsections, we discuss two different methods to allocate power across sub-bands.

DC programming approach
Since the scheduler has the information regarding the users to be scheduled on each sub-band and the fraction of total power that each user should be allocated, the scheduler needs to solve the optimization problem (3.9a) for power variable on each sub-band $P_n \ \forall \ n$. The equivalent of Problem (3.9a) can be written as

$$\text{minimize} \qquad \sum_{n=1}^{N} \sum_{i=1}^{M_r} -w_i^n R_i^n\left(P_n, \bar{\beta}_n\right) \tag{3.15a}$$

$$\text{subject to} \qquad \sum_{n=1}^{N} P_n = P_{max}, \tag{3.15b}$$

$$P_n \geq 0. \tag{3.15c}$$

Note that the above objective function can be decomposed into difference two convex functions:

$$
\begin{aligned}
\sum_{n=1}^{N} \sum_{i=1}^{M_r} & -w_i^n \log_2 \left(\frac{1 + \sum_{k=1}^{i} \bar{\beta}_k^n P_n \Gamma_i^n}{1 + \sum_{j<i} \bar{\beta}_j^n P_n \Gamma_i^n} \right) \\
= \sum_{n=1}^{N} \sum_{i=1}^{M_r} & -w_i^n \log_2 \left(1 + \sum_{k=1}^{i} \bar{\beta}_k^n P_n \Gamma_i^n \right) \\
- \sum_{n=1}^{N} \sum_{i=1}^{M_r} & -w_i^n \log_2 \left(1 + \sum_{j<i} \bar{\beta}_j^n P_n \Gamma_i^n \right) \\
= \mathbf{R}(\mathbf{P}) & - \mathbf{S}(\mathbf{P}),
\end{aligned}
\tag{3.16}
$$

where $\mathbf{P} = \begin{bmatrix} P_1 & P_2 & \cdots & P_n & \cdots P_N \end{bmatrix}^{\mathrm{T}}$. In order to prove that the objective function of Problem (3.16) is a difference of convex functions, we need to show that $\mathbf{R}(\mathbf{P})$ and $\mathbf{S}(\mathbf{P})$ are convex functions. Since both the functions are negative weighted sum of logarithm functions, they are convex functions [Boy08]. The algorithm to solve different convex functions is presented in Algorithm 3.1.

Equal Power Across Sub-bands

Instead of allocating optimized power values across the sub-bands, equal power can be allocated to each sub-band which would result in reduction in both complexity and signaling overhead resulting in reduction in signaling overhead.

Algorithm for difference of convex programming

Problems (3.11) and (3.15) are not convex in general. However, since the objective function is the difference of two convex functions, it can be efficiently solved using numerical methods to get local and at times global optimum solution [AT05, Boy08]. Using successive convex approximation approach, an efficient sub-optimal solution can be found by iteratively solving a sequence of convex sub-problems. The convex sub-problems are obtained by linearizing the non-convex part of the objective function. Let $\mathbf{Q}(\mathbf{x})$ be an objective function which can be expressed as the difference of two convex functions $\mathbf{R}(\mathbf{x})$ and $\mathbf{S}(\mathbf{x})$, i.e., $\mathbf{Q}(\mathbf{x}) = \mathbf{R}(\mathbf{x}) - \mathbf{S}(\mathbf{x})$. Then the convex sub problems are obtained by replacing $\mathbf{S}(\mathbf{x})$ with its first-order approximation at point $\mathbf{x}^{(k)}$, i.e., $\mathbf{S}(\mathbf{x}) = \mathbf{S}(\mathbf{x}^{(k)}) + \nabla \mathbf{S}(x^{(k)})^T (\mathbf{x} - \mathbf{x}^{(k)})$. The generic algorithm to solve optimization problems involving difference of convex functions is given in Algorithm 3.1.

Algorithm 3.1: Iterative, suboptimal solution for DC problems

---

Data: Initial Point $(\mathbf{x}^{(0)})$, Max Iterations (K_{max}), Tolerance (ϵ),
Objective function $\mathbf{Q}(\mathbf{x})$, Convex functions $\mathbf{R}(\mathbf{x})$ and $\mathbf{S}(\mathbf{x})$.
Result: Desired Suboptimal Solution $\mathbf{x}_{subOpt}$.

[1] Set $k = 0$;

[2] Repeat Step (3) to (5) until $|\mathbf{Q}(\mathbf{x}^{(k+1)}) - \mathbf{Q}(\mathbf{x}^{(k)})| \leq \epsilon$ or $k > K_{max}$;

[3] Convex Approximation of $\mathbf{Q}(\mathbf{x})$ at $\mathbf{x}^{(k)}$:

$$\mathbf{Q}^{(k)}(\mathbf{x}) = \mathbf{R}(\mathbf{x}) - \mathbf{S}(\mathbf{x}^{(k)}) - \nabla\mathbf{S}(x^{(k)})^T(\mathbf{x} - \mathbf{x}^{(k)}) \qquad (3.17)$$

;

[4] Solve:

$$\mathbf{x}^{(k)} = \underset{\mathbf{x} \in \chi}{\text{argmin}} \, \mathbf{Q}^{(k)}(\mathbf{x}) \qquad (3.18)$$

;

[5] $k \leftarrow k + 1$.

In (3.17), $\mathbf{Q}^{(k)}(\mathbf{x})$ is convex w.r.t. $\mathbf{x}$. In (3.18), the minimization is performed over the set of feasible solutions (χ). If the constraint set (χ) is compact and continuous, by Cauchy's theorem, the sequence $\{\mathbf{Q}^{(k)}(\mathbf{x})\}$ always converges. The iterative process terminates at no solution improvement with some tolerance limit. Furthermore, if $\mathbf{R}(\mathbf{x})$ and $\mathbf{S}(\mathbf{x})$ are continuously differentiable on the constraint set, Algorithm 3.1 always returns a stationary point of the objective function $\mathbf{Q}^{(k)}(\mathbf{x})$ [VSS10]. However, the attainment of a stationary point does not guarantee global optimal solution. The convex subproblem in the above algorithm, i.e., (3.18) can be solved using standard convex minimization algorithms such as interior point method, sequential quadratic programming, etc.

Modified versions of the optimal algorithm

Some modification can be made in the optimal scheme to allow a restricted number of users without increasing the overall resource allocation complexity by much. Two heuristic algorithms which are modified versions of the optimal scheme to accommodate the restricted number of users on each PRB are presented here.

Heuristic Resource Allocation Algorithms: After the completion of optimal user selection and power allocation process, sub-bands having more than two number of users are marked for second round of user selection and power allocation process to schedule only two users. The users having the highest and lowest CINRs are selected for scheduling. The intuition behind this criteria for user selection lies in the observation given in [LG01] which states that users with diverse channel conditions yield more benefit using SC with SIC compared to similar channel conditions. Post user selection and power allocation can be made in the following ways.

- **Optimal-Heuristic-1**: As per this scheme, the total power across the considered sub-band is divided among the multiplexed users using the existing power ratio for optimal power allocation step.
- **Optimal-Heuristic-2**: In this scheme, the power ratio value is determined using the DC programming approach defined in Section 3.1.1.

Explanation of Schemes Used for Comparison

The following two resource allocation schemes which are available in literature are also used for comparison with the proposed schemes.

- OFDMA with unequal power across sub-bands: In this scheme, the power across sub-bands is determined using the multi-level iterative water filling algorithm described in [Zou09], and each sub-band is assigned to one user only.
- Optimal with equal power across sub-bands: In this scheme, presented in [Shaq09], equal power across sub-bands is considered with no restriction on the number of user that can be multiplexed over a sub-band.

3.1.2 Performance Analysis

In this section, we study the performance of the proposed algorithms through extensive Monte Carlo simulation. A 19-cell hexagonal cellular model with unity frequency reuse has been considered. The inter-site distance is considered to be 500 m. Users with a mean velocity of 3 km/h are uniformly distributed within the center cell. The minimum distance of the users from the base station is kept to be 35 m. Power delay profile of Urban Macro non-line of sight (UMa-NLoS) [IR08] is considered to generate the small scale variation in the channel. ICI is generated assuming that the neighboring cells are operating at full load condition, i.e., all the sub-bands are active throughout the simulation. Furthermore, in case of optimal, DC-DC and

Table 3.1   System parameters

Parameters	Values
Network layout	Hexagonal grid with 19 sites, three sectors per cell
Distance-dependent path loss	133.6+35 $\log_{10}(d_{km})$
eNB transmit power	41 dBm
System bandwidth	4.82 MHz
Subcarrier spacing	15 KHz
Number of sub-bands	12 (24 subcarriers/SB)
TTI duration	1 ms (12 OFDM symbols)
Antenna configuration	SISO
Antenna tilt	12 degree
Rx noise figure	7 dB
Shadow fading standard deviation (dB)	8 dB
Throughput calculation	Based on Shannon's formula (max 6 b/s/Hz)

FTPA-DC schemes where power variations occur in each sub-band, a pre-generated per TTI-based power profile is used for all the interfering base stations. However, since this work focuses on the performance analysis of the discussed algorithms for a single cell scenario, there is no coordination among base stations to mitigate the effect of ICI. The rest of the system parameters is presented in Table 3.1. In order to average SINR across the subcarriers within a sub-band capacity-based SINR averaging[1] method is used. Full buffer best effort traffic model for the simulation purpose is used. For simplicity, ideal interference cancellation is assumed in SIC detection. In addition, OFDMA results are obtained using proportional fair scheduling, which is a natural outcome of the weighted instantaneous sum rate maximization that we have considered in our problem.

The optimal scheme

Figure 3.1(a) gives the distribution of downlink user throughput for OFDMA and the optimal scheme when 10 users are in the system. It is observed that both cell edge throughput and mean cell throughput have increased considerably for the optimal scheme compared to OFDMA. The increase in the mean cell throughput is 12.44% and cell edge throughput is 20.54%. Further, the geometric average user throughput has increased by 12.27% for the optimal scheme compared to OFDMA scheme (refer Table 3.3).

---

[1]$B_{sb}\log(1 + \text{SINR}_{sb}) = B_{sc}\sum_{i=1}^{N_{sc}} \log(1 + \text{SINR}_{sc}^{i})$ where SINR_{sb} is the effective SINR over the sub-band and SINR_{sc}^{i} is the SINR over the ith subcarrier within the sub-band.

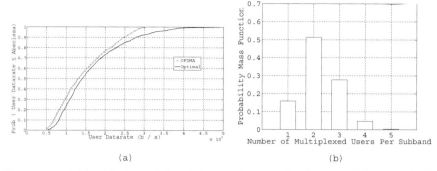

(a)   (b)

Figure 3.1 (a) Distribution of downlink user data rate for the case of 10 users in the system. (b) Number of multiplexed users per sub-band for the case of 10 users in the system. (c) Distribution of downlink post scheduling user SINR with maximum two users per sub-band case.

Figure 3.1(b) shows the probability mass function of number of users multiplexed over a sub-band for the optimal scheme when the system has 10 number of users. It is evident that, by limiting the number of users to be multiplexed over a sub-band to three, a good percentage of gain, achieved through the optimal scheme, can be obtained.

Power allocation schemes

The comparison of following four power allocation schemes along with the greedy user selection scheme is done in this section. The results are generated for the case of 10 users in the system. The legends used for different cases are given in Table 3.2.

The downlink user throughput distributions for FTPA-DC and DC-DC are compared with OFDMA and the optimal scheme in Figure 3.2(a). As expected, the gain obtained by DC-DC is more than all other schemes and comparable with respect to the optimal scheme. A zoomed version of user throughput distribution till 20-percentile is presented in Figure 3.2(b) for all

Table 3.2 Legends used for different power allocation schemes

Algorithm	User Power Assignment Within a Sub-band	Power Allocation Across the Sub-band
FTPA-Equal	FTPA	Equal power allocation
FTPA-DC	FTPA	Iterative DC algorithm
DC-Equal	Iterative DC algorithm	Equal power allocation
DC-DC	Iterative DC algorithm	Iterative DC algorithm

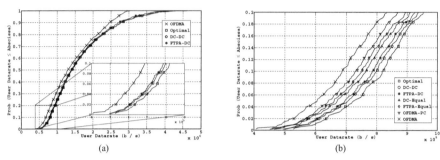

Figure 3.2   (a) Distribution of downlink user data rate for the discussed schemes for the case of 10 users in the system. Distribution curves of FTPA-DC and DC-Equal lies in between DC-DC and FTPA-Equal (not shown in the figure). (b) Distribution of user data rate till 20-percentile point to compare the performance of the discussed schemes for cell edge users. FTPA-DC performs better than DC-Equal scheme bringing more fairness among the users.

the schemes. One important observation is that FTPA-DC is better than DC-Equal in the cell edge region. This gain in the cell edge region for FTPA-DC over DC-Equal comes due to the fact that FTPA-DC performs power allocation across sub-bands which maximizes the system weighted sum rate bringing more fairness among the users in the system.

Table 3.3 presents mean system throughput, mean weighted sum rate, geometric mean throughput, and 5-percentile user throughput for all the schemes for the case of 10 users in the system. The percentage gain of each scheme with respect to OFDMA is presented within parentheses per each of the KPIs. It is clear from Figures 3.2(a) and (b), and Table 3.3 that among the proposed power allocation schemes, DC-DC power allocation scheme gives the best performance followed by DC-Equal, FTPA-DC, and FTPA-Equal. In addition, it can be inferred from Table 3.3 that in case of 10 users in the system and with maximum two users per sub-band scheduling, the gain achieved by DC-DC scheme is 99.56% of the optimal scheme. Moreover, power allocation among multiplexed users through DC programming has clear benefits over the FTPA-based power allocation scheme. One main reason behind this increase in the gain is due to more percentage of user multiplexing in DC programming scheme compared to FTPA scheme as shown in Figure 3.3(a). For FTPA-based power allocation scheme among multiplexed users, the user pairing stays around 60% compared to 90% in case of DC programming-based power allocation scheme. However, the gains obtained

Table 3.3 Comparison of KPIs for different schemes. The percentage of gain achieved over OFDMA by different schemes is given within parentheses

Power Allocation Schemes	Mean System Throughput (Mbps)	Mean Weighted Sum Rate	Geometric Mean Throughput (Mbps)	5-Percentile User Throughput (kbps)
OFDMA	14.53	1.0122e+03	1.3035	602.2
OFDMA-PC	14.70	1.0470e+03	1.3309	646.6
FTPA-Equal (Max 2 Users/SB)	15.788 (8.66%)	1.0860e+03 (7.29%)	1.4031 (7.64%)	660.2 (9.63%)
FTPA-Equal (Max 3 Users/SB)	16.008 (10.17%)	1.1020e+03 (8.87%)	1.4216 (9.06%)	673.5 (11.84%)
FTPA-DC (Max 2 Users/SB)	15.944 (9.73%)	1.1162e+03 (10.27%)	1.4285 (9.59%)	696.8 (15.71%)
DC-Equal (Max 2 Users/SB)	16.204 (11.52%)	1.1167e+03 (10.32%)	1.4408 (10.53%)	681.2 (13.12%)
DC-Equal (Max 3 Users/SB)	16.287 (12.09%)	1.1241e+03 (11.06%)	1.4478 (11.07%)	689.7 (14.53%)
DC-DC (Max 2 Users/SB)	16.268 (11.96%)	1.1354e+03 (12.17%)	1.4543 (11.57%)	707.2 (17.44%)
Optimal-Equal	16.301 (12.44%)	1.1323e+03 (13.39%)	1.4541 (12.27%)	701.9 (20.54%)
Optimal	16.338 (12.44%)	1.1477e+03 (13.39%)	1.4635 (12.27%)	725.9 (20.54%)
Optimal-Heuristic-1	15.968 (9.90%)	1.1327e+03 (11.90%)	1.4338 (10.0%)	739.2 (22.75%)
Optimal-Heuristic-2	16.301 (12.19%)	1.1323e+03 (11.87%)	1.4541 (11.55%)	739.9 (16.56%)

for power allocation across sub-bands through the DC programming exhibit only a marginal improvement in the mean system throughput compared to the equal power distribution scheme. In addition, the performances of Optimal-Heuristic-1 and Optimal-Heuristic-2 are comparable to DC-DC and optimal scheme in terms of mean system throughput. Moreover, both the heuristic algorithms perform better than all the schemes in terms of 5-percentile user throughput.

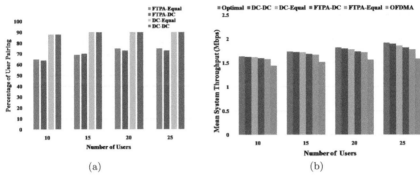

Figure 3.3   (a) Percentage of times two users are multiplexed over a sub-band for different power allocation schemes with respect to different numbers of users in the system. Owing to higher percentage of user multiplexing, higher gain is observed in case of DC-Equal and DC-DC. (b) Mean system throughput as a function of number of users.

Multi-user Diversity: The effect of multi-user diversity gain on mean system throughput is given in Figure 3.3(b). It is clear that increasing the number of users in the system improves the mean system throughput for all the mentioned schemes. Moreover, the relative gain for each of the schemes with respect to OFDMA improves as the number of users increases in the system.

3.2 Coordinated OFDM – NOMA

Modern multi-carrier cellular systems, such as LTE and WiMAX, reuse the entire available spectrum in every cell, which gives rise to ICI. ICI plays a dominant role in limiting the potential performance benefits that would have been, otherwise, achieved in the absence or weak interference scenario. The adverse effect of ICI is more prominent for cell edge users than cell center users as they receive less desired power from the serving base station and more interference power from the interfering base stations. Cooperation (coordination) among base stations, so that the cell edge throughput of each base station is improved with minimal reduction or if possible improvement in mean cell throughput, is an approach to address ICI. In this section, the possible benefit that can be achieved through coordination among base stations for a multi-carrier NOMA system is explored.

Coordination for OFDM-NOMA systems, where individual base stations can multiplex multiple users on a single sub-band, has been addressed in [Cho14]. In [Har14], authors have proposed a dynamic FFR scheme for

OFDM-NOMA systems with the objective of satisfying a target cell edge user throughput while improving the overall system throughput.

In this part, a centralized coordinated scheduling scheme, where all the coordinating base stations are attached to a centralized node that possesses complete channel state information (CSI) of all the users in the network, is described. The motivations behind proposing a centralized coordination scheme are as follows:

1. The proposed centralized scheme can be considered for implementation in a C-RAN type cellular architecture.

2. The solution obtained from the centralized scheme can be used as a benchmark for different distributed resource allocation algorithms for multi-cellular OFDM-NOMA systems.

3.2.1 System Model

The performance of a downlink OFDM-NOMA-based cellular system with K number of coordinating base stations is described here. A simple scenario where each base station is serving equal number of user M is considered. The total bandwidth of the system (BW_{sys}) is equally partitioned into N number of orthogonal sub-bands with bandwidth BW_{sb}. Moreover, each sub-band consists of equal number of subcarriers in the frequency domain and spans over one transmission time interval (T_{tti}) in the time domain. As we have considered non-orthogonal resource sharing among users attached to a particular base station, a base station can simultaneously schedule more than one user on a sub-band.

The signal model for the multicell scenario is similar to that of a single cell scenario, except for the fact that the ICI is explicitly considered. Considering that all the users, which are attached to a single base station, can be multiplexed on the nth sub-band of the kth base station (SB_k^n), the superposition coded symbol that is transmitted by the base station on each subcarrier of SB_k^n is given as

$$x_k^n = \sum_{m=1}^{M} x_{m,k}^n \sqrt{P_{m,k}^n},$$

(3.19)

where $x_{m,k}^n$ is the complex QAM symbols intended for the mth multiplexed users on n^{th} sub-band of kth base station ($\text{UE}_{m,k}^n$). Moreover, the QAM symbols are drawn from a constellation with unit mean and unit variance.

$P^n_{m,k}$ is the power allocated to the user $UE^n_{m,k}$ and P^n_k is the total power on SB^n_k honoring the equality $\sum_{m=1}^{M} P^n_{m,k} = P^n_k$. It is to be noted that in contrast to the single-cell system model, here, the concept of power ratio is dropped and the power values for each multiplexed user are considered explicitly.

The symbol received by $UE^n_{m,k}$ on $SB^n_{m,k}$ is

$$y^n_{m,k} = h^{n,k}_{m,k} x^n_k + \sum_{l=1,l\neq k}^{K} h^{n,l}_{m,k} x^n_l P^n_l + v^n_{m,k} \tag{3.20a}$$

$$= \underbrace{\sqrt{P^n_{m,k}} h^{n,k}_{m,k} x^n_{m,k}}_{\text{desired signal}} + \underbrace{\sum_{i=1,i\neq k}^{M} \sqrt{P^n_{i,k}} h^{n,k}_{m,k} x^n_{i,k}}_{\text{intra cell interference}} + \underbrace{\sum_{l=1,l\neq k}^{K} h^{n,l}_{m,k} x^n_l P^n_l + w^n_m}_{\text{inter cell interference}}, \tag{3.20b}$$

where $h^{n,l}_{m,k}$ is the complex channel response between $UE^n_{m,k}$ and the lth base station. It captures the effect of path loss, shadowing, and small scale fading. $w^n_{m,k}$ constitutes of both additive Gaussian white noise with zero mean and variance $\sigma^2_{n,sb}$, i.e., average noise power over a sub-band and uncoordinated ICI.

We define channel gain to inter-cell interference plus noise ratio (CINR) as

$$\Gamma^{n,l}_{m,k} = \frac{|h^{n,k}_{m,k}|^2}{\mathcal{N}^n_{m,k} + \sum_{k=1}^{K} P^n_l |h^{n,l}_{m,k}|^2}, \tag{3.21}$$

where $\mathcal{N}^n_{m,k}$ captures the power due to noise and uncoordinated interference. The user experiencing a better CINR will always be able to remove the interference received from the symbols belonging to the user with poorer channel condition and the user experiencing the worst channel condition treats the interference from the rest of the users as noise during decoding process. Hence, if the channel gains of a set of users multiplexed over SB^n_k are order as $\Gamma^{n,k}_{1,k} \geq \Gamma^{n,k}_{2,k} \geq ... \geq \Gamma^{n,k}_{m,k} \geq \Gamma^{n,k}_{m+1,k} \geq ... \geq \Gamma^{n,k}_{M,k}$, then post SIC SINR of $UE^n_{m,k}$ on SB^n_k is given as

$$SINR^n_{m,k}(\mathbf{P}_n) = \frac{P^n_{m,k} G^{n,k}_{m,k}}{1 + \sum_{i=1}^{m-1} G^{n,k}_{m,k} P^n_{i,k} + \sum_{l=1,l\neq k}^{K} G^{n,l}_{m,k} P^n_l}, \tag{3.22}$$

where $G_{m,k}^{n,l} = \frac{|h_{m,k}^{n,l}|^2}{\mathcal{N}_{m,k}^n}$ represents the channel gain to noise ratio (CNR) of $UE_{m,k}^n$ on SB_k^n and $\mathbf{P}_n \in \mathcal{R}^{KM \times 1}$ is a vector containing the power values of $P_{m,k}^n$ for $m = 1, 2, \dots, M$ and $k = 1, 2, \dots, K$.

For the case when two users are multiplexed over a sub-band, the SINR of individual users on the sub-band is given by

$$\text{SINR}_{1,k}^n(\mathbf{P}_n) = \frac{P_{1,k}^n G_{1,k}^{n,k}}{1 + \sum_{l=1, l \neq k}^{K} G_{m,k}^{n,l} P_l^n}, \tag{3.23a}$$

$$\text{SINR}_{2,k}^n(\mathbf{P}_n) = \frac{P_{2,k}^n G_{2,k}^{n,k}}{1 + G_{1,k}^{n,k} P_{1,k}^n + \sum_{l=1, l \neq k}^{K} G_{m,k}^{n,l} P_l^n}, \tag{3.23b}$$

where (3.23a) represents the SINR of the user who is experiencing a better channel condition and performs SIC to remove intra-cell interference, and (3.23b) represents the SINR of the user who is experiencing poorer channel condition and treats the intra-cell interference as noise during the decoding process.

Using the expressions for SINR in (3.23a) and (3.23b), the total number of bits that can be transmitted to $UE_{m,k}^n$ can be written as

$$R_{m,k}^n(\mathbf{P}_n) = SB_{BW} T_{tti} \log_2(1 + \text{SINR}_{m,k}^n(\mathbf{P}_n)), \tag{3.24}$$

where SB_{BW} is the bandwidth of a sub-band and T_{tti} is the span of a sub-band in the time domain.

It is assumed that all the coordinating base stations are connected to a centralized scheduling entity and the central scheduler possesses complete CSI of all the users attached to the coordinating base stations. The job of the central scheduler is to determine: (1) the set of co-channel users to be scheduled on each resource block for each base station and (2) the power that needs to be allocated to each multiplexed user with the objective of maximizing the weighted sum rate of the system subject to per base station power constraint. In order to reduce the signaling overhead, it is considered that a maximum two users can be multiplexed on each sub-band. The optimization problem to be solved by the scheduling unit is given as

$$\underset{P_{m,k}^n \geq 0}{\text{maximize}} \quad \sum_{k=1}^{K} \sum_{n=1}^{N} \sum_{m=1}^{M} \alpha_{m,k}^n w_{m,k}^n R_{m,k}^n(\mathbf{P}_n) \tag{3.25a}$$

$$\text{subject to} \qquad \sum_{n=1}^{N}\sum_{m=1}^{M} P_{m,k}^{n} \leq P_{\max} \quad k = 1, 2, \ldots, K, \qquad (3.25b)$$

$$\sum_{m=1}^{M} \alpha_{m,k}^{n} \leq 2, \qquad (3.25c)$$

where $w_{m,k}^{n}$ is the weight of $UE_{m,k}^{n}$ and $P_{\max}$ is the maximum available power for each base station. Furthermore, $\alpha_{m,k}^{n}$ is the binary random variable defined as

$$\alpha_{m,k}^{n} = \begin{cases} 1 & \text{if } UE_{m,k}^{n} \text{ is assigned to } SB_{m,k}^{n} \\ 0 & \text{Otherwise} \end{cases} \qquad (3.26)$$

As we have considered proportional fair scheduling, the weights are reciprocal of the long-term average data rate of each user. In that sense, the utility function that we have considered is equivalent to proportional fair utility function [JPP00].

3.2.2 Solution Methodology and Proposed Algorithms

In order to solve Problem (3.25), we proceed in two steps. In the first step, the scheduler decides the user pair that should be multiplexed over each sub-band. In the second step, given the information on the pair of users on each sub-band, the scheduler allocates power to each user.

User pair selection and power allocation initialization

1. Greedy user selection with DC-Equal power allocation: In this user selection step, the DC-Equal algorithm described in the previous chapter along with the greedy user selection scheme is used to select the pair of users to be scheduled on each sub-band.

   For the following three algorithms, first the optimal algorithm is run for individual BSs to select the set of users to be multiplexed on each sub-band.

2. Max CINR difference from optimal solution: From the set of multi-plexed users, two users with maximum CINR difference are selected for each sub-band.

3. Best WSR and worst CINR from optimal solution: From the set of multiplexed users given by the optimal solution, the user that provides

the best WSR and the user with the worst CINR are selected for further consideration. The power allocation on each sub-band remains the same as given by the optimal algorithm.

4. Pair of users giving maximum sum WSR: From the set of multiplexed users given by the optimal solution, the user pair providing maximum sum weight sum rate is selected. The power allocation on each sub-band remains the same as given by the optimal algorithm.

The optimal user selection scheme is used in every base station to select the set of co-channel users that gives the maximum weighted sum rate. Since we have considered the case where maximum two users can share the sub-band within a base station, from the optimal user set, we select the two users with maximum CINR difference. The steps determine the values that $\alpha_{m,k}^n$ should take.

Power allocation

For a give pair of selected users, the power allocation problem can be framed as

$$\underset{P_{m,k}^n \geq 0}{\text{maximize}} \quad \sum_{k=1}^{K} \sum_{n=1}^{N} \sum_{m=1}^{2} w_{m,k}^n R_{m,k}^n(\mathbf{P}_n) \tag{3.27a}$$

$$\text{subject to} \quad \sum_{n=1}^{N} \sum_{m=1}^{2} P_{m,k}^n \leq P_{\max} \quad \text{for all } k = 1, 2, \ldots, K, \tag{3.27b}$$

where with a slight abuse of notation, we define the variables $w_{1,k}^n$ and $R_{1,k}^n$ as the weight and rate of the user experiencing better channel condition on SB_k^n; and $w_{2,k}^n$ and $R_{2,k}^n$ correspond to the weight and rate of the user experiencing poorer channel condition on SB_k^n. Moreover, $\mathbf{P}_n$ now belongs to $\mathbb{R}^{2N \times 1}$ with elements $P_{1,k}^n$ corresponding to the power allocated to the user experiencing better channel condition and $P_{2,k}^n$ corresponding to the power allocated to the user experiencing poorer channel condition on SB_k^n.

For a given set of selected users, the set of optimal power values must satisfy the first-order KKT conditions [Boy08]. The partial Lagrangian of Problem (3.27) can be written as

$$L(\mathbf{P}_n, \boldsymbol{\lambda}) \triangleq \sum_{k=1}^{K} \sum_{n=1}^{N} \sum_{m=1}^{2} w_{m,k}^n R_{m,k}^n(\mathbf{P}_n) - \boldsymbol{\lambda}^{\mathrm{T}} \left(\sum_{n=1}^{N} \sum_{m=1}^{2} P_{m,k}^n - P_{\max} \right),$$

$$\tag{3.28}$$

where the variable $\mathbf{P} \in \mathcal{R}^{2KN \times 1}$ is a vector containing the power values of each multiplexed on each sub-band of each base station, i.e., $\mathbf{P} = \begin{bmatrix} \mathbf{P}_1^T & \cdots & \mathbf{P}_n^T & \cdots & \mathbf{P}_N^T \end{bmatrix}$, and $\lambda \in \mathcal{R}^K$ is the vector containing Lagrangian multipliers for each base station. The optimum values of the elements in $\mathbf{P}$ and λ must satisfy the following equalities:

$$\frac{\partial L}{\partial P_{m,k}^n} \triangleq 0$$

$$\Rightarrow \frac{w_{m,k}^n G_{m,k}^{n,k}}{1 + \sum_{j=1}^m P_{j,k}^n G_{m,k}^{n,k} + \sum_{l=1,l\neq k}^K P_l^n G_{m,k}^{n,l}} = \log(2)\lambda_k + t_{m,k}^n, \quad (3.29a)$$

$$\lambda_k \left(\sum_{n=1}^N \sum_{m=1}^2 P_{m,k}^n - P_{\max} \right) = 0, \quad (3.29b)$$

$$\lambda_k \geq 0, \quad (3.29c)$$

$$\sum_{n=1}^N \sum_{m=1}^2 P_{m,k}^n \leq P_{\max}, \text{ and } P_{m,k}^n \geq 0 \quad (3.29d)$$

for $k = 1, \ldots, K$; $n = 1, \ldots, N$; and $m = 1, 2$. Equation (3.29a) reflects the stationarity condition for optimal value, (3.29b) represents the complementary slackness condition, (3.29c) provides the dual feasibility constraint, and (3.29d) ensures the primal feasibility. Also,

$$t_{m,k}^n \triangleq \sum_{j=m+1}^2 \frac{w_{j,k}^n \text{SINR}_{j,k}^n G_{j,k}^{n,k}}{\log_e(2) \left[1 + \sum_{i=1}^j P_{i,m}^n G_{i,m}^{n,m} + \sum_{l=1,l\neq k}^K P_l^n G_{i,m}^{n,l} \right]}$$

$$\underbrace{\qquad\qquad\qquad\qquad\qquad\qquad\qquad\qquad\qquad\qquad\qquad}_{\text{Intra cell component}}$$

$$+ \sum_{q=1,q\neq k}^K \sum_{j=1}^2 \frac{w_{j,q}^n \text{SINR}_{j,q}^n G_{j,q}^{n,k}}{\log_e(2) \left[1 + \sum_{i=1}^j P_{i,q}^n G_{i,q}^{n,q} + \sum_{l=1,l\neq q}^K P_l^n G_{i,q}^{n,l} \right]}.$$

$$\underbrace{\qquad\qquad\qquad\qquad\qquad\qquad\qquad\qquad\qquad\qquad\qquad}_{\text{inter-cell component}}$$

$$(3.30)$$

The following modified water-filling algorithm is used to find at least a local optimal value $P_{m,k}^n$ given the set of multiplexed users. From (3.29a), given a value of Lagrangian multiplier (λ_k) for the kth base station, $P_{m,k}^n$, i.e., the power for the mth multiplexed user on the nth sub-band of the kth base station is given as

$$P_{m,k}^n = \left[\frac{w_{m,k}^n}{\log_e(2)\lambda_k + t_{m,k}^n} - \frac{1 + \sum_{j=1}^{m-1} P_{j,k}^n G_{m,k}^{n,k} + \sum_{l=1,l\neq k}^{K} P_l^n G_{k,m}^{n,l}}{G_{m,k}^{n,k}} \right]^+,$$

(3.31)

where $[x]^+ = \max(0,x)$. In (3.31), λ_k along with $t_{m,k}^n$ dictates the water-filling level.

Next to find λ_k, the following equality needs to be solved.

$$\sum_{n=1}^{N} \sum_{m=1}^{2} P_{m,k}^n = P_{\max}.$$

(3.32)

For a fixed $t_{m,k}^n$, (3.32) is a monotonic function of λ_k. Hence, (3.32) can be efficiently solved by a one-dimensional search method such as bisection method. If $\sum_{n=1}^{N} \sum_{m=1}^{2} P_{m,k}^n < P_{\max}$, λ_k should be decreased. Otherwise, λ_k should be increased. The above process goes on until (3.32) is satisfied. If no positive value of λ_k satisfies (3.32), then it is set to zero. In this case, the kth base station does not use the total available power for transmission. The summary of the proposed algorithm is given in Algorithm 3.2.

Algorithm 3.2: Multicell NOMA iterative power allocation algorithm

Data: Max Iterations ($L_{\max}$), Set $l = 0$

[1] Select the sets of users to be multiplexed on each sub-band of each base station and initialize the dual variables **P** and $\boldsymbol{\lambda}$ as per Section 3.2.2;

[2] Compute $t_{m,k}^n \forall m, k, n$ as per (3.30);

[3] **while** *No Convergence of WSR or $l < L_{max}$* **do**

[4]   **while P** *has not converged* **do**

[5]   **for** $k = 1$ *to* K **do**

[6]   Update $P_{m,k}^n \forall m, n$ according to (3.31)

[7]   **end**

[8]   **end**

[9]   Update $\boldsymbol{\lambda}$ as per (3.32) and tax term $t_{m,k}^n \forall m, \forall k, \forall n$ as per (3.30)

[10] **end**

Performance evaluation environment and assumptions

A hexagonal cellular model with 27 base stations (sites) using unity frequency reuse as shown in Figure 3.4 is considered. Each site is divided into three sectors. The inter-site distance is considered to be 500 m. Coordination is considered among sector-1, sector-2, and sector-3 respectively. The rest of the sectors contribute to the uncoordinated interference. Users with mean velocity 3 km/h are uniformly distributed in each coordinating sector with a minimum distance of 25 m and a maximum distance of 290 m. In addition, it is assumed that each sector is having equal number of users.

The power received at the mth user attached to the kth sector from the lth sector is calculated using the pathloss formula

$$\text{PL}_{m,k}^{l}(d) = 133.6 + 35\log_{10}\left(\frac{d_{m,k}^{l}}{1000}\right) + \chi \ \text{dB}, \tag{3.33}$$

where $d_{m,k}^{l}$ is the distance in meters between the mth user attached the kth sector and the lth sector and χ is a normal random variable with zero mean and σ_{sh} standard deviation. To capture the effect of small scale fading, the power delay profile of non-line of sight scenario of the sub-urban macro environment is considered [IR08].

In Figure 3.4, except sectors 1, 2, and 3, the rest of the sectors belong to the set of uncoordinated interferes ($\mathcal{I}$). We have only considered large-scale power contribution from the uncoordinated interferers, which makes the interference from uncoordinated interferers static. Moreover, we vary the level of interference received from uncoordinated interferers, to verify the robustness of the proposed algorithms. The uncoordinated static interference is treated as noise at the users; hence, the effective noise variance at the mth user attached to the kth base station on the nth sub-band would be

$$\mathcal{N}_{m,k}^{n} = \sigma_{n,\text{sb}}^{2} + \sum_{i \in \mathcal{I}} 10^{\frac{-\text{PL}_{m,k}^{i}}{10}} \frac{P_{\max}}{N\Delta}, \tag{3.34}$$

where $\sigma_{n,\text{sb}}^{2}$ is the actual noise variance on the nth sub-band. The arguments behind the consideration of variable uncoordinated interference are:

1. To quantify the effect of cluster size: In case of weak uncoordinated interference, the coordinating base stations are the major source of interference to each other. However, in case of strong uncoordinated interference, there are high chances that the major source of interference does

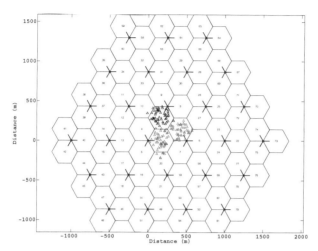

Figure 3.4 Simulated network with 27 cells (81 sectors). Users are distributed in the sector numbered 1, 2, and 3.

not belong to the set of coordinating base stations. Hence, analyzing the system performance, in both these cases, will help us quantify the performance loss due to the limited size of the clusters.

2. To quantify the performance in a noise-limited environment: In case of strong uncoordinated interference, the system becomes noise limited. Hence, the performance benefits that the proposed algorithm provides in case of a noise-limited scenario can also be evaluated.

Depending on the scenarios mentioned above, the SINR across subcarriers for each user is found during every scheduling interval. In order to average SINR across the subcarriers within a sub-band, a capacity-based averaging method is used. The throughput is calculated using Shannon's formula. We have assumed that there is no feedback delay between the users and base stations, and base stations and the central scheduler. In addition, we have considered a full buffer best effort traffic model for simulation purpose. The rest of the system parameters are presented in Table 3.4.

Key performance indicators (KPIs)

In Table 3.5, we summarize the baseline algorithms with which we compare the proposed schemes. We have used three kinds of performance metrics, namely, mean cell throughput, cell edge user throughput, per TTI instantaneous weighted sum rate, and geometric average user of throughput (GAT)

Table 3.4   System parameters

Parameters	Values
Network layout	Three coordinating sectors facing each other in cloverleaf fashion
Distance-dependent pathloss	$133.6 + 35 \log_{10}(d_{km})$
eNB transmit power	46 dBm
System bandwidth	4.82 MHz
Subcarrier spacing	15 kHz
Number of useful subcarriers	288
Number of useful subcarriers per sub-band	24
Number of sub-bands	12
TTI duration	1 ms (12 OFDM symbols)
SINR averaging across subcarriers	capacity-based Appendix II
Directional antenna gain	17 dB
Antenna configuration	SISO
Antenna gain pattern	Appendix I
Receiver noise figure	7 dB
Shadow fading standard deviation (dB)	8 dB
Throughput calculation	Based on Shannon's formula

Table 3.5   Description of baseline algorithms used for comparison purposes

Schemes	User Selection	Power Allocation Algorithm	Comments
SC-MLWF	Max WSR	Multilevel iterative waterfilling	OFDMA based No coordination
MC-IIWF	Max WSR	Multi-cell improved iterative waterfilling	OFDMA based Centralized coordination
SC-DC-DC	Greedy user selection	DC-DC	NOMA based No coordination

to evaluate the performance of various algorithms discussed earlier. The motivation behind using the GAT is that maximizing GAT is the objective of our per TTI weighted sum rate maximization problem. In addition, we have considered the 5-percentile point of user throughput distribution curve as the measure of cell edge user throughput.

In order to normalize the performance of different algorithms, during a single drop, each algorithm operates on the same user location, shadowing profile, and small scale variation in the channel. A single drop runs for 1000 TTIs and the averaging window for evaluating the weights is taken as 100 ms. The result for the average user throughput is calculated at the end of each drop. Furthermore, for better averaging purpose, 64 such drops are performed with different user locations, shadowing profiles, and small-scale channel variations. The results for the KPIs such as post-scheduling SINR, mean system throughput, geometric average user throughput, and 5-percentile cell edge user throughput are collected after 64 drops.

Results
Weak uncoordinated interference

We begin the evaluation of the performance of the schemes, with respect to the metrics per TTI (instantaneous) weighted sum rate (IWSR) and GAT, which we have targeted to maximize in the framed optimization problem. Columns 1 and 2 of Table 3.6 give the achieved mean IWSR and GAT for various schemes. The improvement in IWSR for NOMA with coordination over NOMA without coordination is 5.14% for weak uncoordinated interference. This improvement in weak uncoordinated interference results in 3.64% improvement in GAT. For the comparison purpose, we observe the same trend for OFDMA with and without coordination cases. An interesting observation worth mentioning is that NOMA without coordination using the DC-DC power allocation algorithm performs better than OFDMA with coordinated scheduling in terms of both ISWR and GAT.

Next we compare other important KPIs such as mean system throughput and 5-percentile cell edge user throughput for the uncoordinated and coordinated system. The user throughput distribution curve is presented in Figure 3.5(a). Though user throughput distributions for OFDMA with and without coordination are also presented, out main interest lies in comparing the performance of NOMA with and without coordination. The mean system throughput and 5-percentile point of user throughput distribution (also referred as the cell edge user throughput) are presented in Table 3.6. NOMA with coordination provides an improvement of 14.5% and 1.52% in the cell edge user throughput and mean cell throughput.

Strong uncoordinated interference

In the strong uncoordinated interference scenario, coordination among sectors gives improvement for IWSR and GAT

Figure 3.5 presents the distribution of downlink user datarate for the proposed resource allocation scheme, both for weak and strong uncoordinated interference scenario. It is clear from the figure that for both the scenarios, there is a considerable amount of improvement in the cell edge user throughput. While the improvement in cell edge user throughput for weak uncoordinated interference is 14.56%, for strong uncoordinated interference,

Table 3.6 Comparison of KPIs in case of weak uncoordinated interference

Schemes	Mean Weighted Sum Rate	Geometric Average Throughput (bps)	Mean Cell Throughput (bps)	5-Percentile Throughput (bps)	Convergence (%)
OFDMA (MLWF)	669.8252	2.65E+06	2.86E+07	1.32E+06	-
OFDMA (MC-IIWF)	740.4405	2.85E+06	3.03E+07	1.58E+06	98
NOMA (SC-DC-DC)	778.7227	3.03E+06	3.29E+07	1.58E+06	-
NOMA (MC-NOMA-IIWF)	818.2177	3.13E+06	3.34E+07	1.81E+06	87.5

Table 3.7 Comparison of KPIs in case of strong uncoordinated interference

Schemes	Mean Weighted Sum Rate	Geometric Average Throughput (bps)	Mean Cell Throughput (bps)	5-Percentile Throughput (bps)	Convergence (%)
OFDMA (MLWF)	377.1477	1.53E+06	1.69E+07	7.25E+05	-
OFDMA (MC-IIWF)	384.7444	1.55E+06	1.71E+07	7.43E+05	100
NOMA (SC-DC-DC)	403	1.62E+06	1.80E+07	7.77E+05	-
NOMA (MC-NOMA-IIWF)	422.6124	1.69E+06	1.86E+07	8.46E+05	99

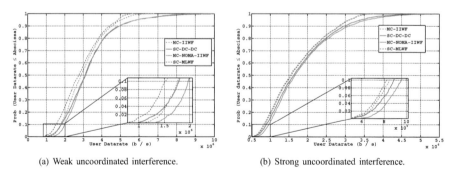

(a) Weak uncoordinated interference.

(b) Strong uncoordinated interference.

Figure 3.5 Distribution of downlink average user data rate.

the value drops to 8.9%. In contrast, the improvement in mean cell throughput for strong uncoordinated interference is 4.32% compared to 1.52% for weak uncoordinated interference. It can be inferred that if the cluster size in case of strong uncoordinated interference becomes larger, the improvement in cell edge user throughput can be more.

3.3 User Selection and Optimal Power Allocation in NOMA

In this part a low-complexity user set selection and optimal power allocation for weighted sum rate maximization for NOMA is presented. The described method achieves performance similar to [OKH12] but at a much lower complexity. A closed-form solution for optimal power allocation in cellular downlink is discussed.

3.3.1 System Model

Let us consider a hexagonal cellular layout with unity frequency reuse, where each cell has K users distributed uniformly in it. Total bandwidth of system B_{sys} is equally divided into N number of sub-bands(SB), each having a bandwidth B_{sb}. Equal power distribution across sub-bands is considered. Thus, total power for each sub-band can be written as $P = P_{sys}/N$. For K users multiplexed on the nth SB (SB_n), the superposition coded signal transmitted from base station can be written as:

$$x_n = \sum_{i=1}^{K} \sqrt{p_i^n} x_i^n \qquad (3.35)$$

where x_i^n are complex QAM symbols drawn from a constellation with zero mean and unit variance, and p_i^n is the power allocated to the ith user such that $\sum_{i=1}^{K} p_i^n = P$. If the ith user is not scheduled on SB_n, then $p_i^n = 0$. The received signal by the ith user can be written as:

$$y_i^n = h_i^n x_n + v_i^n \qquad (3.36)$$

where h_i^n is the complex channel response between the base station and the ith user on SB_n inclusive of the effects of small and large scale fading (shadowing as well as path loss). The term v_i^n constitutes additive white Gaussian noise (AWGN) and ICI. Users will use SIC receiver to reduce inter-user interference faced by them, thus, increasing received SINR. According to the principle of SIC, optimal order for decoding is in the increasing order of

CINR given by $\Gamma_i^n = \frac{|h_i^n|^2}{E[|v_i^n|^2]}$ [Y. 13]. Thus, for user set sorted in decreasing order of CINR ($|\Gamma_1^n| \geq \cdots \geq |\Gamma_K^n|$), user having the ith index will not face any interference from users having index i+1, i+2,... K. Hence, received SINR post SIC for the ith user in SB_n can be written as

$$\gamma_i^{n,post} = \frac{p_i^n|h_i^n|^2}{\sum_{j=1}^{i-1} p_j^n|h_i^n|^2 + E[|v_i^n|^2]} = \frac{p_i^n\Gamma_i^n}{1 + \sum_{j=1}^{i-1} p_j^n\Gamma_i^n} \tag{3.37}$$

Using (3.37), throughput R_i^n and spectral efficiency SE_i^n of the ith user within the nth sub-band becomes

$$R_i^n = B_{sb}\log_2(1 + \gamma_i^{n,post}), SE_i^n = \log_2(1 + \gamma_i^{n,post}). \tag{3.38}$$

Average data rate T_i^n of the ith user within the nth sub-band at time (t+1) is defined [OKH12, Eq. (6)], [SDS15, Eq. (9)] as: $T_i^n(t+1) \triangleq \{1-\frac{1}{t_c}\}T_i^n(t) + \frac{1}{t_c}R_i^n(t)$, where t denotes the time index representing a subframe index and t_c defines the time horizon for throughput averaging. $R_i^n(t)$ is the throughput of the ith user at time instance t. $R_i^n(t)$ is calculated using Shannon's capacity formula (3.38) and is zero if the ith user is not scheduled.

The optimization problem to be solved can be represented as:

$$\text{Maximize} \quad R_{sum} = \sum_{i=1}^{K} w_i^n SE_i^n$$

$$\text{subject to} \sum_{i=1}^{K} p_i^n = P, \& \ p_i^n \geq 0, \ \forall \ i = 1, 2, ..K, \tag{3.39}$$

where w_i^n is the weight associated with the ith user over the nth sub-band. The weight is considered as the reciprocal of past average data rate $T_i^n(t)$ of the ith user [OKH12]. The past average data rate is assumed to remain constant between the time interval t and $(t + 1)$ and the user selection and power allocation solution is to be executed in each interval. Thus, for brevity, the time index "t" is dropped.

Power Allocation
The solution to above problem may be arrived at using the Lagrangian multiplier method. The Lagrangian function can be expressed as:

$$L = -\sum_{i=1}^{K} w_i^n SE_i^n - \sum_{i=1}^{K} \lambda_i p_i^n + \nu\{\sum_{i=1}^{K} p_i^n - P\} \tag{3.40}$$

where λ_i and ν are the Lagrange multipliers corresponding to the constraints $p_i^n \geq 0$ and $\sum_{i=1}^{K} p_i^n = P$, respectively, such that $\lambda_i \geq 0$. Each user belonging to the chosen user set will take part in the power allocation process for optimization problem (3.39). Let us assume that we solve the optimal power allocation problem for the chosen user set U_{opt} having M number of users where $M \leq K$. Thus, constraints $p_i \geq 0$ and corresponding lagrange multiplier λ_i need not be considered. The solution can be obtained by taking derivative of lagrangian function L with respect to p_i^n. The generalized solutions can be written as

$$1 + \sum_{j \leq i} \Gamma_i^n p_j^n = \frac{w_i^n \Gamma_i^n}{\nu + \sum_{j=i+1}^{M} x_j}, \ \forall \ i = M, M-1, M-2, ...1$$

(3.41)

where $x_j = \frac{\{\nu + \sum_{k=j+1}^{M} x_k\}^2}{\frac{w_j^n}{p_j^n} - \nu - \sum_{k=j+1}^{M} x_k}$. Now, we have M power variables and one

Lagrange multiplier ν which can be solved using above M equations and total power constraint $\sum_{i=1}^{M} p_i^n = P$. Thus, we obtained a closed-form expression for power allocation among users belonging to U_{opt} as

$$\nu = \frac{w_M^n \Gamma_M^n}{1 + P\Gamma_M^n}, \quad p_1^n = P - \sum_{i=2}^{M} p_i^n$$

$$p_i^n = \frac{\frac{w_i^n - w_{i-1}^n}{\nu + \sum_{j=i+1}^{M} x_j} - \frac{1}{\Gamma_i^n} + \frac{1}{\Gamma_{i-1}^n}}{1 - \frac{w_{i-1}^n}{w_i^n}}$$

(3.42)

$$\forall \ i = M, M-1, M-2, ...2,$$

where $x_j = \frac{\{\nu + \sum_{k=j+1}^{M} x_k\}^2}{\frac{w_j^n}{p_j^n} - \nu - \sum_{k=j+1}^{M} x_k}$. Thus, starting from the Mth user, optimal

power can be allocated to users in first attempt/single iteration by substituting the values of known variables in the above derived solution.

User Selection
Let us consider that user set S has users sorted in decreasing order of CINR. Here, the generalized conditions that have to be satisfied for any two consecutive users to belong in the sorted user set (U_{opt}) are described.

It can be shown that if (3.43) is satisfied, then the ith user will not belong to U_{opt}.

$$\frac{w_{i-1}^n}{w_i^n} > \frac{\Gamma_i^n\{1 + \Gamma_{i-1}^n\{P - \sum_{j=i+1}^{K} p_j^n\}\}}{\Gamma_{i-1}^n\{1 + \Gamma_i^n\{P - \sum_{j=i+1}^{K} p_j^n\}\}} \tag{3.43}$$

Similarly, the $(i-1)^{\text{th}}$ user will not belong to U_{opt}. if,

$$\frac{w_{i-1}^n}{w_i^n} < \frac{\Gamma_i^n}{\Gamma_{i-1}^n} \tag{3.44}$$

Both i^{th} and $(i-1)^{\text{th}}$ users will belong to U_{opt} if and only if it satisfies

$$\frac{\Gamma_i^n}{\Gamma_{i-1}^n} < \frac{w_{i-1}^n}{w_i^n} < \frac{\Gamma_i^n\{1 + \Gamma_{i-1}^n\{P - \sum_{j=i+1}^{K} p_j^n\}\}}{\Gamma_{i-1}^n\{1 + \Gamma_i^n\{P - \sum_{j=i+1}^{K} p_j^n\}\}} \tag{3.45}$$

3.3.2 Performance Analysis

A 19-cell hexagonal model with unity frequency reuse is considered with system parameters as described in previous sections. Power delay profile of sub-urban non-line of sight is considered to generate the small scale variation in the channel. ICI is generated by considering the neighboring cells operating at full load. Full buffer best effort traffic model is considered for evaluation.

In Figure 3.6(a), geometric mean user throughput for different power allocation schemes has been shown. It can be seen that the above-described

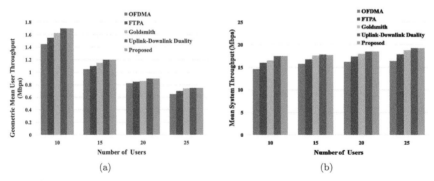

Figure 3.6   (a) Geometric mean user throughput with respect to the number of users (b) Mean system throughput with respect to the number of users.

scheme achieves a maximum geometric mean user throughput. Geometric mean user throughput decreases with an increase in the number of users because the total system bandwidth and total power in each sub-band are constant. In terms of overall system performance as well (Figure 3.6(b)), the performance of the above-described scheme is at par with optimal uplink–downlink duality scheme and greater than other schemes.

4

Millimeter-Wave Communication

With the world moving from fourth-generation (4G) to fifth-generation (5G) mobile standards, the researchers [Yos] feel that the huge data requirements in 5G ($\sim$1000 times that of 4G [A$^+$14]) cannot be met by only increasing the spectral efficiency (SE) (by using single or multi-user (MU) multiple-input multiple-output (MIMO) technology [FG98, Tel99, GKH$^+$07]) of 4G networks. Some of the other methods suggested in the literature [A$^+$14] are extremely aggressive frequency reuse by using very dense deployment of network nodes and use of larger bandwidths. It is likely that the final solution would be a combination of all of these methods. So search for unused bandwidth to support wireless Internet with datarates of several Gbps is expected to remain relevant in the coming years. In the recent World Radio Conference (WRC) in 2015, some of the sub-6-GHz spectra which were discussed for application in 5G were around 470–694 MHz, 1427–1518 MHz, 3.3–3.8 GHz, and 4.5–4.99 GHz [H$^+$16a]. Apart from this, one of the decisions taken in WRC is new study efforts on several spectral bands above 24 GHz for the next WRC in 2019, which will provide cornerstone for development of 5G services. In this chapter, we present a brief survey of the mmwave channel models and some aspects of millimeter-wave (mmW) communications.

Some of the identified frequency bands for mmW communication are around 28 GHz, 38 GHz [RSM$^+$13a], 60 GHz [XKR02, YXVG11], 73 GHz, and E-band (mainly from 71 to 76 GHz and from 81 to 86 GHz) [WLSV15]. Each of these bands have their own advantages and disadvantages [Yos]. It has been already predicted [RSM$^+$13a, WLSV15] that, the major difference between 5G and 4G standards would be the use of the mmW spectrum. Wireless communication is also looking to be a substitute for the multi-gigabit fiber-optic communication systems specially in the places where laying optical fiber networks are cost ineffective [EDC14, EDC15]. Rapid progress in complementary metal oxide semiconductors (CMOS) and radio frequency (RF) integrated circuit technologies [DES+04, GAPR09, RMG11] has also encouraged this move to shift to higher frequencies. IEEE has already recommended standards for mmW wireless communication technologies at 60 GHz, such as IEEE 802.15.3c for wireless personal area networks (WPANs) [Yon07], and IEEE 802.11ad for wireless local area networks (WLAN) [agi13] and ECMA-387 [ET08, AMS13]. Technologies to enhance SE, like Massive MIMO [RPL+13], where a huge number of antennas are put together at the BS have an inherent advantage in mmW due to smaller element size.

4.1 Millimeter-Wave Channel Modeling

The properties which set these frequencies apart from the conventional sub-6-GHz frequency bands are as below.

- First, as per Friis transmission formula [Fri46], the free space pathloss (FSPL) is proportional to the square of the carrier frequency, so the amount of path loss (PL) at 60 GHz is 20 dB higher than that of 6 GHz. We need high gain antenna and beamforming (BF) technologies [KS16] to compensate for this loss. As a result, mmW communications are inherently directional, sometimes called *quasi-optical*. This makes such links inherently immune to interference at the cost of increased transceiver complexity.
- Second, the atmospheric and rain attenuations are more severe in mmW frequencies. Absorption due to atmospheric oxygen can go as high up to 15 dB/Km [GLR99] at frequencies around 60 GHz. The disadvantage of this is that this huge atmospheric absorption makes 60 GHz communication short ranged; however, this short range gives inherent security from eavesdroppers, which, in turn, can be thought as an advantage.
- Third, due to a combination of low beamwidth antennas and weak diffractional ability, mmW signals experience blockage by humans and other objects [MMS+09, MML+11, J+13, tg3]. Also, at these frequencies, specular reflections dominate diffused reflection components and most multipath components (MPCs) are specular reflected, also several MPCs tend to have similar time and angular characteristics so such MPCs bunched together to form a *cluster*. Clustering can be identified as visual inspection [SV87, SJJS00] to complex processes like kernel density estimation [C+03]. A comprehensive study of different clustering methods can be found in [Gus14]. The line-of-sight (LOS) ray carries

the most power, where as even the strongest non-line-of-sight (NLOS) clusters are much weaker than the LOS ray [BAH14]. Therefore, most of the power propagated between the transmitter (Tx) and the receiver (Rx) is through the LOS ray (if present) and a few low-order NLOS clusters [MML$^+$11].

- Fourth, polarization [JB68] plays an important path in mmW. It has been seen that the power loss between the Tx and the Rx due to polarization mismatch can be as high as 20 dB [MMS$^+$09, MPM$^+$10, MML$^+$11].

The success of both high SE technologies and extreme network densification depends on the accuracy wireless channel model used. This becomes even more important in mmW, where channels are less "random" due to the presence of LOS rays and specular reflection becoming dominant. For sub-6 GHz frequencies, we have mature channel models starting from the historical Okumura model [OOF68], Hata model [Hat80], and the COST-231-Walfisch-Ikegami model [WB88] to the more recent models such as Third Generation Partnership Project Spatial Channel Model (3GPP-SCM) [3gp03], its extension (SCME) [BSG$^+$05], and WINNER II [K$^+$07]. A comparison between these models can be found in [NST$^+$07]. However, as we have seen, the mmW channel is much different as compared to that of at sub-6 GHz, so the sub-6-GHz models have to be discarded or at least modified to be use in the mmW. The METIS [N$^+$15] project has in-depth modeling of the same parameters as in WINNER II, such as delay distribution, angle of arrival (AoA) distribution, angle of departure (AoD) distribution, and correlation between these parameters related to various scenarios such as dense urban, urban, rural, indoor and highway, BS to user equipment (UE), backhaul (BH) (BS-BS link), device to device (D2D), and vehicle to vehicle (V2V) for outdoor to outdoor (O2O), outdoor to indoor (O2I), and indoor to indoor (I2I), topologies for both sub-6 GHz and mmW frequency ranges. There have been several research papers touching various aspects of mmW channel modeling as well as survey papers like [ZOY14, SJK$^+$03, Han15] for channel modeling for different frequencies and [KS16, RMG11, NLJ$^+$15, GLR99] considering different aspects of mmW communication.

4.1.1 Radio and Propagation Channel Models

The local propagation scenario of a channel is called the *propagation channel*, and is given as $h(t, \tau, \phi_{Tx}, \theta_{Tx}, \phi_{Rx}, \theta_{Rx})$, describes the double-directional radio channel, which depends only on the propagation scenario and the frequency of operation and is independent of Tx and Rx radiation patterns, where t is the absolute time, τ is the arrival time of clusters and rays, ϕ_{Tx} is the azimuth AoD, θ_{Tx} is the elevation AoD, ϕ_{Rx} is the azimuth AoA, and θ_{Rx} is the elevation AoA [SMB01].

Once we have calculated $h(t, \tau, \phi_{Tx}, \theta_{Tx}, \phi_{Rx}, \theta_{Rx})$, we can find the channel impulse response (CIR) by integrating with Tx and Rx antenna radiation patterns as shown below,

$$h(t, \tau, \phi_{Rx}, \theta_{Rx})$$
$$= \int_{\theta_{Tx}=0}^{\pi} \int_{\phi_{Tx}=0}^{2\pi} h(t, \tau, \phi_{Tx}, \theta_{Tx}, \phi_{Rx}, \theta_{Rx}) G_T(\phi_{Tx}, \theta_{Tx}) d\theta_{Tx} d\phi_{Tx}$$

(4.1)

where $G_T(\phi_{Tx}, \theta_{Tx})$ describes the Tx antenna radiation pattern. The single-directional radio channel described by $h(t, \tau, \phi_{Rx}, \theta_{Rx})$ is dependent on the propagation scenario, frequency of operation, and Tx antenna radiation pattern but is independent of Rx antenna radiation pattern. Finally, we can integrate the product of $h(t, \tau, \phi_{Rx}, \theta_{Rx})$ and $G_R(\phi_{Rx}, \theta_{Rx})$ to get the radio channel which depends on the propagation scenario, frequency of operation, and both Tx and Rx antenna radiation patterns. Where $G_R(\phi_{Rx}, \theta_{Rx})$ is the Rx antenna radiation pattern.

$$h(t, \tau) = \int_{\theta_{Rx}=0}^{\pi} \int_{\phi_{Rx}=0}^{2\pi} h(t, \tau, \phi_{Rx}, \theta_{Rx}) G_R(\phi_{Rx}, \theta_{Rx}) d\theta_{Rx} d\phi_{Rx} \quad (4.2)$$

4.1.2 Large-Scale Channel Model

The two major characteristics of large-scale channel models are PL and shadowing. PL is the ratio of power received at Rx to the power transmitted at Tx. The well-known Friis [Fri46] equation for Friis' FSPL, in dB, is given by

$$FSPL = 20log_{10}(\frac{4\pi f_c d}{c}) \quad (4.3)$$

In the above equation, f_c is the center frequency of transmission, d is the distance between Tx and Rx, and c is the speed of light. From (4.3), it is clear that FSPL is directly proportional to the square of f_c, so FSPL at higher frequencies (like in mmW bands) is much higher than in lower frequencies (sub-6 GHz bands). Although (4.3) is simple, it suffers from a major limitation that it assumes a single transmission path between Tx and Rx, which is very often not the case in real life. It is well known [Gol05,

Rap02] that if there is a reflected ray along with the direct ray, then for large Tx–Rx separation PL varies with d^4 instead of d^2 as predicted by (4.3). In general, it is often assumed that the PL varies with d^n, where n is called PL exponent (PLE).

Apart from the deterministic part, there is a random part in the path loss, called shadowing, which usually models the additional path loss due to the presence of large objects such as hills, buildings, etc., in the propagation path. In dB scale, the large-scale PL is a normal distributed random variable, which, with an assumption that $f_c \gg$ signal bandwidth, can be written as [YXVG11]

$$PL(d) = \overline{PL}(d) + X_\sigma \qquad (4.4)$$

In the above equation, $\overline{PL}$ is the average PL [YXVG11, Smu09], in dB, and X_σ is the large-scale fading, which is often depicted as log-normal distribution in linear scale [Gol05], consequently which becomes normal in logarithmic scale. $\overline{PL}$, in dB, is represented as,

$$\overline{PL}(d) = PL(d_0) + 10n log_{10}(d/d_0) + \sum_{q=1}^{Q} X_q \qquad (4.5)$$

where d_0 is the reference distance (often taken as 1 metre), where the communication is LOS, n is the PLE, and X_q is the additional attenuation by the blocking effect of the qth object.

PL, at a Tx–Rx separation of d, can also be calculated by taking [Smu09]

$$PL(d) = P_{Tx} - \overline{P}_{Rx}(d) + G_{Tx} + G_{Rx} \qquad (4.6)$$

where P_{Tx} is the transmitted power (in dB), $\bar{P}_{Rx}(d)$ is the average received power at Rx, and G_{Tx} and G_{Rx} are the Tx and Rx antenna gains-respectively (in dBi). The averaging is done to average out the small-scale variations. Also, it is assumed that the Tx and Rx antennas are perfectly aligned and are LOS. An effect of removing the antenna effects is the influence of the environment may get lost, which can be minimized by using omnidirectional antennas [Smu09]. However, directional antennas are necessary for a more stable link budget, so values of both n and variance of X_σ are given for different types of antennas of varying gains.

Rain attenuation usually increases with frequencies. This has been seen in measurements taken by International Telecommunication Union (ITU) [itu]. However, Qingling [QL06] has shown that, for a Tx and Rx distance of 230 m and using a f_c of 35 GHz, the rain attenuation remains under 20 dB in 99.9%

cases, and under 10 dB for approximately 99.5% of the cases. Rappaport [RSM+13a] in his work has used Qingling's value and shown that for a heavy rainfall of 25 mm/h, at a Tx–Rx distance of 200 m, the rain attenuation is just 1.4 dB, at 38 GHz, it is around 1.6 dB, and although not specifically mentioned, at 60 GHz, it is about 2 dB. The detailed values of rain attenuation are given in Figure 4.1.

The amount of atmospheric attenuation varies greatly with frequency in the 25 to 100 GHz frequency range. This is shown in Figure 4.2. This

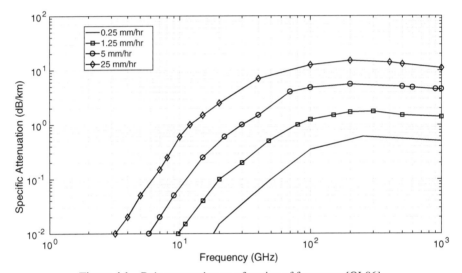

Figure 4.1   Rain attenuation as a function of frequency [QL06].

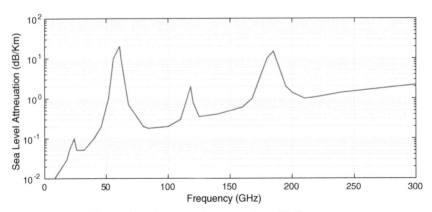

Figure 4.2   Atmospheric attenuation with frequency.

shows that some frequencies like 28 and 38 GHz have very low values of atmospheric attenuation, so they are quite good for longer range communications; however, around 60 GHz, the value of attenuation is quite high, so this frequency can only be used for very short ranged communication and will give automatic security and high frequency reuse. Gianetti *et al.* in their survey paper [GLR99] have analyzed these effects in detail.

Considering the above effects, from (4.4) and (4.5), ignoring attenuation by objects, we can write [GLR99]

$$PL(d) = PL(d_0) + 10n log_{10}(d/d_0) + X_\sigma + \alpha_{tot}(d - d_0)/1000 \quad (4.7)$$

where α_{tot} is the total atmospheric attenuation (in dB/Km). This has two components given by

$$\alpha_{tot} = \alpha_{rain} + \alpha_{atmosphere} \quad (4.8)$$

where α_{rain} is the attenuation due to rain and $\alpha_{atmosphere}$ is the attenuation due to atmosphere (rain and water vapor). If γ is the PLE taking including these attenuations, then by simple algebraic manipulations of (4.4), (4.5), and (4.7), and assuming $d \gg d_0$, we write

$$\gamma = n + \frac{1 \times 10^{-4} \times \alpha_{tot} \times d}{log_{10}(d/d_0)} \quad (4.9)$$

Assuming a cell size of 500 m, [GLR99] shows that the carrier to interference ratio (C/I) at 60 GHz (with α_{oxygen} = 15 dB/Km and a frequency reuse factor of 4) is almost 15 dB higher than that of at 900 MHz at cell boundary. [PvL93] derives the expressions for CIR in the following cases: i) desired signal is LOS (Rice distributed) and interference is NLOS (Rayleigh distributed) found in microcellular environments, ii) both desired signal and interference are LOS (found in picocellular environments), and also iii) where both are NLOS were considered, and it was shown that the outage probability is dramatically reduced in the presence of O_2 absorption.

Some works like [ALS$^+$13] prefer using a slightly modified version of (4.4) which fits better with the measured results

$$PL(d) [dB] = \alpha + 10\beta log_{10}(d) + \xi \quad (4.10)$$

where d is the Tx–Rx distance (in meters), α and β are the least square fit of floating point intercept and slope over the measured distances, and ξ is the shadow fading which is normal distributed with mean 0 and variance σ^2.

This is called the floating intercept path loss model and it does not come from propagation geometry. This model tends to minimize the shadowing effect.

The METIS channel model [N$^+$15] modifies the ITU-R M.2135 Urban Microcell (UMi) PL model [itu09] as follows,

$$PL(d_1) \, [dB] = 10n_1 log_{10}(d_1) + 28 + 20log_{10}(f_c) + PL_{1|dB} \quad (4.11)$$

The above equation is valid for Tx–Rx distances (in meters) such that 10 m $< d_1 < d_{BP}$, n_1 is the PLE, f_c is the frequency in GHz, and $PL_{1|dB}$ and d_{BP} are given by

$$PL_{1|dB} = -1.38log_{10}(f_c) + 3.34 \quad (4.12)$$

$$\alpha_{BP} = 0.37e^{-\frac{log_{10}(f_c)}{1.13}} \quad (4.13)$$

and

$$d_{BP} = \alpha_{BP}\frac{2\pi h_{BS}h_{UE}}{\lambda} \quad (4.14)$$

In (4.12) and (4.13), f_c is in GHz, h_{BS} and h_{UE} are the BS and UE antenna heights, and λ is the wavelength.

For $d_{BP} < d_1 < 500$ m, (4.11) can be modified as

$$PL(d_1) \, [dB] = 10n_2 log_{10}(d_1) + 7.8$$
$$-18log_{10}(h_{BS}h_{UE}) + 20log_{10}(f_c) + PL_{1|dB} \quad (4.15)$$

f_c is in GHz, and h_{UE} and h_{BS} are in meters.

n_1 is found to be 2.2, $n_2 = 4$ and the log-normal distributed shadow fading standard deviation is found to be 3.1, and these terms exhibit frequency independence.

In IEEE 802.15.3c channel model [YXVG11], the channel PL model is depicted as

$$\bar{PL}(d) = PL(d_0) + 10nlog_{10}(\frac{d}{d_0}), \quad \text{for } d \geq d_0 \quad (4.16)$$

where the symbols have the same significance as in (4.5). The Tx and Rx antenna gains can be compensated by using adjusting the proposed value of parameter $PL(d_0)$ by $G_{Tx} + G_{Rx}$ [PSML06].

For IEEE 802.11ad standard, the PL model that is considered is given by [YXVG11]

$$\bar{PL}(d) = A_c + 20log_{10}(f) + 10nlog_{10}(d) \quad (4.17)$$

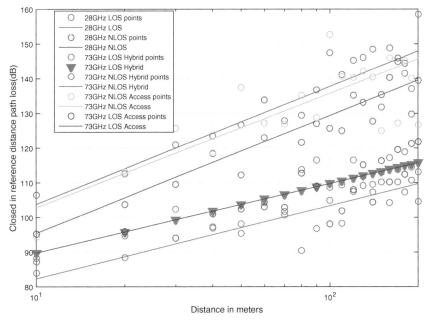

Figure 4.3 Closed in reference distance path losses for different scenarios at two frequencies.

where A_c is a constant, f is 60 GHz, n is the PLE, and d is the Tx–Rx distance.

To give a comparison among the different PL models, the geometrical PL model as given in (4.5) has been plotted in Figure 4.3 for two different mmW frequencies 28 and 73 GHz. The frequency dependence of PL is clearly visible. The floating intercept PL models have also been compared in Figure 4.4. We see that the floating intercept PLs generally have a lesser slope and also have higher values than the physical propagation path loss models as evident in Figure 4.5.

Small-Scale Channel Model

The main feature of propagation at mmW wavelengths is the so-called "quasi-optical" nature of the rays due to higher frequencies [MMS+09]. As a result, the diffraction effects are smaller; hence, shadow loss due to objects is significantly larger as mentioned in (4.5). The penetration loss is also higher at these frequencies [Z+13]. If high-gain Rx antennas are used, then the received signal consists of the LOS ray and clusters of reflected rays each

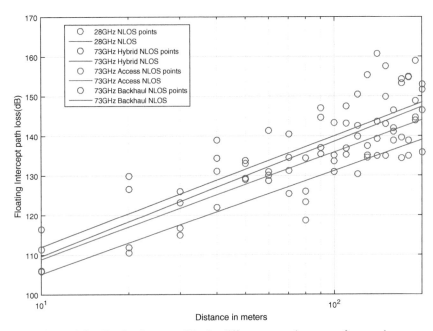

Figure 4.4   Floating intercept PLs for different scenarios at two frequencies.

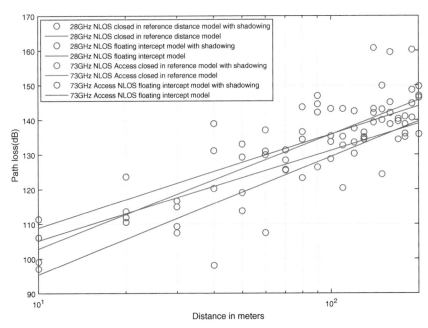

Figure 4.5   Comparison of closed in reference point and floating intercept PL models.

having AoA and AoD very close to the LOS ray. Experimental investigations [MMS$^+$09] also suggest that a reflected path usually consists of several rays having very close time of arrival (ToA), AoA, and AoD [MML$^+$11]. Such a collection of rays having similar properties is called a "cluster" [SV87].

As an example, the small-scale channel model for IEEE 802.15.3c standard is defined using the Saleh–Valenzuela [SV87, YXVG11] equation given by

$$h(t,\phi,\theta) = \beta\delta(\tau,\phi,\theta)$$
$$+ \sum_{l=0}^{L}\sum_{k=0}^{K_l} \alpha_{k,l}\delta(t - T_l - \tau_{k,l})\delta(\phi - \Omega_l - \omega_{k,l})\delta(\theta - \Psi_l - \psi_{k,l})$$

$$(4.18)$$

In the above equation, the first term, $\beta\delta(\tau,\phi,\theta)$, accounts for the gain of the LOS component, β is a complex number denoting the magnitude and phase of the LOS component, l denotes the lth cluster, L is the total number of clusters, k denotes the kth ray, K_l denotes the total number of rays in the lth cluster, T_l, Ω_L, and Ψ_l denotes the mean ToA, AoA (both planes), and AoD (both planes), respectively, of the lth cluster, and $\alpha_{k,l}$, $\tau_{k,l}$, $\omega_{k,l}$, and $\psi_{k,l}$ denote the amplitude and phase, ToA, AoA, and AoD, respectively, of the kth ray of the lth cluster [YXVG11]. The complex numbers β and $\alpha_{k,l}$ are modeled in polar form as

$$\alpha_{k,l} = \beta_{k,l}e^{j\theta_{k,l}} \qquad (4.19)$$

where $\beta_{k,l}$ are statistically independent, positive, random variables whose probability distributions are discussed in Section 4.1.2 and $\theta_{k,l}$ is uniformly distributed between 0 and 2π. It is important to note that (4.18) gives a *double-directional channel model* as discussed earlier. In order to get the impulse response of the multipath channel, we have to take the Tx and Rx antenna radiation patterns into account as [A$^+$07]

$$h(t,\tau) = \int_{\phi=0}^{2\pi}\int_{\theta=0}^{\pi} h(t,\phi,\theta)G_{Tx}(\phi)G_{Rx}(\theta)f(\tau - \tau')d\phi d\theta d\tau' \quad (4.20)$$

where $h(t,\phi,\theta)$ can be found from (4.18), $G_{Tx}(\phi)$ and $G_{Rx}(\theta)$ are the radiation patterns of Tx and Rx antennas respectively, and $f(\tau - \tau')$ is the overall impulse response of the Tx and Rx antennas and filters.

In a different approach to channel modeling, [TNMR14b] improves the mmW channel modeling done in [RRE14, ALS+13] at 28 and 73 GHz. The authors fill up the measurement gaps at these frequencies by ray tracing and suggest a correlated set of random variables similar to the 3GPP model [TR14]. It uses a set of seven correlated Gaussian random variables $x(1)$ to $x(7)$ to model different parameters like ToA, AoA, AoD, etc. We now explain the important small scale channel parameters one by one.

LOS Blockage Probability

As already mentioned, mmW is extremely susceptible to be blocked by humans, objects, building, etc., due to its poor diffraction properties. So often the LOS component is blocked resulting in a sharp decrease of the received power. In IEEE 802.11ad channel modeling the probability of blockage is set to either 0 or 1 as a model parameter [YXVG11]. Many other works, however, model the blockage as a probabilistic quantity. For example, in [BAH14], a stochastic building model approach is described where the building locations and parameters are described by probability distributions. Using Boolean scheme of rectangles, and assuming that the centers of these rectangles are distributed as a Poisson Point Process, it is given [BH13] that the LOS blockage probability is given as

$$p(R) = e^{-\beta R} \qquad (4.21)$$

where R is the distance between Tx and Rx and β is a parameter depending on the density and average size of the buildings. Alternatively, in [ALS+14], the blockage probability has three states, complete outage, LOS, and NLOS where complete outage denotes no signal received. The expressions for the probabilities of these states are,

$$p_{out}(d) = \max(0, 1 - e^{(-a_{out}d + b_{out})})$$
$$p_{LOS}(d) = (1 - p_{out}(d))e^{-a_{los}d}$$
$$p_{NLOS}(d) = 1 - p_{out}(d) - p_{LOS}(d) \qquad (4.22)$$

where a_{out}, b_{out}, and a_{LOS} are obtained from the measurements. $p_{LOS}(d)$ and $p_{NLOS}(d)$ denote the probabilities of LOS/NLOS at a distance d from the Tx.

NLOS Cluster Blockage Probability

For NLOS clusters, the probability of blockage depends on i) the surrondings (i.e., indoor or outdoor, distance of the reflecting surface etc.), ii) in case of

indoor, whether the reflection takes place from walls, ceiling, or floor, and
iii) the order of the reflection, in case of statistical modeling.

Number of Clusters

In some cases, like 802.11ad, the number of clusters is a deterministic number
[MML+11], some other works, like [TNMR14b], take the same approach.
However, some other authors [M+06] prefer to model the number of clusters
to be Poisson distributed as follows

$$pdf_L(L) = \frac{(\bar{L})^L e^{(-\bar{L})}}{L!} \qquad (4.23)$$

where $x!$ denotes the factorial of x and the mean, $\bar{L}$, varies from 1 to 13.6,
depending on the scenario. At mmW, say at 60 GHz, some authors [YSH07,
GKZV09] even use a single cluster model. For IEEE 802.15.3c, for various
scenarios, it was seen that the number of clusters does not follow any specific
distribution, but it is known to vary between 3 and 14 depending on whether
LOS or NLOS case is considered and the environment [YXVG11].

Arrival Times

The cluster arrival times and ray arrival times are usually denoted by two
Poisson processes [YXVG11]. Under this model the cluster inter-arrival
time and ray inter-arrival times are two independent random variables. In
particular, the cluster arrival time is exponentially distributed random variable
given as [SV87, Shu04]

$$p(T_l|T_{l-1}) = \Lambda e^{-\Lambda(T_l-T_{l-1})}, \ for \ l > 0 \qquad (4.24)$$

where Λ is the cluster arrival rate.

The ray arrival time has a similar expression

$$p(\tau_{k,l}|\tau_{k-1,l}) = \lambda e^{-\lambda(\tau_{k,l}-\tau_{k-1,l})}, \ for \ k > 0 \qquad (4.25)$$

where λ is the ray arrival rate.

A different approach is to methodically develop the channel coefficients
from the parameters given in [K+07, N+15], and [TNMR14b] using equations. One such work is that of Thomas *et al.* [TNMR14b]. Here the cluster
delays are exponentially distributed as

$$\tau_n = -r_\tau \sigma_\tau ln(X_n) \qquad (4.26)$$

where τ_n is the arrival time of the nth cluster. [TNMR14b] lists the delay scaling parameter, $r_\tau = 3$, X_n is a unit variance Gaussian random variable and σ_τ, which is the delay spread, is given as

$$\sigma_\tau = 10^{\epsilon_{DS} x(1) + \mu_{DS}} \tag{4.27}$$

where $\epsilon_{DS} = 0.3$, $\mu_{DS} = -6.8$, and $x(1)$ is the first of the seven correlated Gaussian random variables mentioned earlier. The delays are sorted as per the equation $\tau_n = sort(\tau_n - min(\tau_n))$.

$$\tau_{n,m} = \begin{cases} \tau_n, \ m = 1, 2, 3, 4 \\ \tau_n + \ 5 \ ns, \ m = 5, 6, 7 \\ \tau_n + \ 10 \ ns, \ m = 8, 9, 10 \end{cases} \tag{4.28}$$

Cluster and Ray Power Distribution

As per the original Saleh model [SV87], the cluster power follows exponential distribution as

$$\bar{\beta}_{k,l}^2 = \bar{\beta}^2(T_l, \tau_{k,l}) = \bar{\beta}^2(0, 0) e^{-\frac{T_l}{\Gamma}} e^{-\frac{\tau_{k,l}}{\gamma}} \tag{4.29}$$

where $\beta_{k,l}$ is from (4.19), whose rms values are monotonically decreasing as shown in the above equation. Γ and γ are the power-delay constants for clusters and rays, respectively, and $\bar{\beta}^2(0, 0)$ is the average power gain of the first ray of the first cluster.

In the IEEE 802.15.3c model [YXVG11], the power of the LOS ray, β, is given as

$$\beta[dB] = 20log_{10}[\frac{\mu_d}{d} | \sqrt{G_{t1}G_{r1}} + \sqrt{G_{t2}G_{r2}} \Gamma_0 e^{j\frac{4\pi h_1 h_2}{\lambda d}} |] - PL_d(\mu_d) \tag{4.30}$$

where $PL_d(\mu_d)$ is given by

$$PL_d(\mu_d) = 20log_{10}(\frac{4\pi d_0}{\lambda}) + A_{NLOS} + 10n_d log_{10}(\frac{d}{d_0}) \tag{4.31}$$

where λ, A_{NLOS}, μ_d, Γ_0, and h_1 and h_2 are wavelength, attenuation value for the NLOS environment, mean distance, reflection coefficient and height of Tx and Rx, and G_{t1}, G_{t2}, G_{r1}, and G_{r2} are the gains of Tx antenna for paths 1 and 2 and the gain of Rx antenna for paths 1 and 2. Both cluster and ray amplitudes are modeled by log-normal distribution following the equation,

$$p_l(r) = \frac{1}{\sqrt{2\pi}\sigma_r r} e^{-\frac{(ln(r) - \mu_r)^2}{2\sigma_r^2}} \tag{4.32}$$

where $\mu_r = \mathbb{E}[(r)]$ and σ_r^2 are the mean and variance of the Gaussian $ln(r)$, respectively, and $\mathbb{E}$ denotes the expectation operator.

The approach used for modeling IEEE 802.11ad is relatively simpler [YXVG11], where a Friis model is used to calculate the amplitude of the LOS path, given as

$$\beta_{LOS}[dB] = G_t + G_r + 20log_{10}(\frac{\lambda}{4\pi d}) \quad (4.33)$$

where G_t and G_r the Tx and Rx antenna gains, λ is the wavelength, and d is the distance of the LOS path. The gain of a cluster β_i is given by

$$\beta_i[dB] = 20log_{10}(\frac{g_i\lambda}{4\pi(d+R)}) \quad (4.34)$$

where once again d is the LOS distance, λ is the wavelength, g_i is the reflection loss, and R is the total Tx–Rx distance followed by the ith cluster path reduced by d. g_i is known to follow log-normal distribution with mean -10 dB and variance of 4 dB for the first-order reflection and mean of -16 dB and variance of 5 dB for the second-order reflection [YXVG11] or truncated log-normal distribution [MML+11]. However, [Gus14] argues that, based on the measured data, whether a truncated log-normal approach is reasonable.

In order to model the power of a ray within a cluster, three types of rays are specified in IEEE 802.11ad standard, which are N_{pre} number of *pre-cursor rays*, a single *central ray*, and N_{post} number of *post-cursor rays*. The arrival times of both pre- and post-cursor rays are modeled using Poisson process given in (4.25) with rates of λ_{pre} and λ_{post}, respectively. The average amplitudes and delays of a particular pre-cursor or post-cursor rays decay as $A_{pre}(\tau)$ and $A_{post}(\tau)$ are exponentially decaying at rate γ_{pre} and γ_{post}, respectively. $A_{pre}(0)$ and $A_{post}(0)$ are related to $\alpha_{i,0}$ by,

$$K_{pre} = 20log_{10}(|\frac{\alpha_{i,0}}{A_{pre}(0)}|) \quad (4.35)$$

$$K_{post} = 20log_{10}(|\frac{\alpha_{i,0}}{A_{post}(0)}|) \quad (4.36)$$

where K_{pre} and K_{post} are the K-factors of pre- and post-cursor rays [YXVG11].

As per Thomas *et al.*'s work [TNMR14b], the cluster powers are generated according to the following equation

$$P_n = e^{-\tau_n \frac{r_\tau - 1}{r_\tau \sigma_\tau}} 10^{-0.1Z_n} \quad (4.37)$$

where τ_n, r_τ, and σ_τ can be found from 4.1.2 and Z_n is normally distributed from mean 0 and variance ζ^2 [TNMR14b].

Arrival/Departure Angles

The AoA and AoD distributions are often the same. So only one of the two distributions are reported here, and the other distribution would be similar. The cluster arrival angles are often distributed uniformly between 0 and 2π, given by [YXVG11]

$$p(\Omega_l|\Omega_0) = \frac{1}{2\pi}, \; for \; l > 0 \tag{4.38}$$

On the other hand, the arrival angles of rays within a cluster are given as Gaussian or Laplacian distribution [YXVG11, agi13, iee09, Gus14].

$$p(\omega_{k,l}) = \frac{1}{\sqrt{2\pi}\sigma_\phi} e^{-\frac{\omega_{k,l}^2}{2\sigma_\phi^2}} \tag{4.39}$$

$$p(\omega_{k,l}) = \frac{1}{\sqrt{2}\sigma_\phi} e^{-|\frac{\sqrt{2}\omega_{k,l}}{\sigma_\phi}|} \tag{4.40}$$

where σ_ϕ is the variance.

Some other references like [ALS+14] report that at 28 and 73 GHz, the horizontal (azimuth) cluster arrival angles for both BS and UE are uniformly distributed between 0 and 2π. The BS and UE vertical cluster central angles are reported to be same as LOS elevation angle. The BS cluster rms angular spread is exponentially distributed.

Recent works like [TNMR14b, H+16b] report that the azimuth AoD of the nth cluster is given as

$$\phi_{n,AoD} = X_n\phi_n + Y_n + \phi_{LoS,AoD} \tag{4.41}$$

where $\phi_{LoS,AoD}$ is the LoS azimuth AoD and AoA at the Tx and Rx after their locations are defined in system simulations, and

$$\phi_n = \frac{2\sigma_{AS,D}\sqrt{-1.4ln(P_n/max(P_n))}}{1.4C} \tag{4.42}$$

where $X_n \in \{1, -1\}$ is uniformly distributed random variable, constant C is a scaling factor related to the total number of clusters C is scaled to 0.9 in this model with fixed $N = 6$, and Y_n is a normal distributed random variable with mean 0 and variance $\sigma_{ASD}/7$ [H$^+$16b]. Finally, the ray angles are calculated by random offset angles α_m given by

$$\phi_{n,m,AoD} = \phi_{n,AoD} + \alpha_m \qquad (4.43)$$

where α_m is a Laplacian random variable with zero mean and standard deviation as the intra-cluster RMS azimuth spread of departure.

On the other hand, the AoA and AoD in the zenith (elevation) direction follows Laplacian distribution. The elevation equation follow similar equations as in azimuth.

$$\theta_{n,EoD} = X_n\theta_n + Y_n + \theta_{LoS,EoD} + \mu_{Offset,EoD} \qquad (4.44)$$

where $\theta_{LoS,EoD}$ is the LOS direction in the elevation AoD direction for both Tx and Rx.

$$\theta_n = \frac{2\sigma_{ES,D}\sqrt{-1.4ln(P_n/max(P_n))}}{C} \qquad (4.45)$$

where a scaling factor C is given as 0.98. The elevation spread of departure $\sigma_{ES,D}$ is an exponential random variable characterized by $\lambda_{ES,D}$, which is given by

$$\frac{1}{\lambda_{ES_D}(d)} = max(\gamma_1 d_{2D} + \eta_1, \gamma_2 d_{2D} + \eta_2) \qquad (4.46)$$

where d_{2D} is the 2D distance between Tx and Rx and γ and η are given in [H$^+$16b].

The offset EoD angle is modeled by

$$\mu_{Offset,EoD}(d_{2D}) = -10^{(a_{EoD}log_{10}(max(b_{EoD},d_{2D}))+C_{EoD})} \qquad (4.47)$$

The values of a_{EoD}, b_{EoD}, and C_{EoD} can also be found from [H$^+$16b]. The subpath EoD angle is given as

$$\theta_{n,m,EoD} = \theta_{n,EoD} + \alpha_m \qquad (4.48)$$

Power Delay Spectrum

The power DS (PDS) of the channel, also known as power delay profile (PDP). The PDP of the channel at a time instant t can be given by [YXVG11]

$$A_c(t,\tau) = \mathbb{E}[|h(t,\tau)|^2] \qquad (4.49)$$

If we assume that the channel coherence time is high, then we can drop the time-dependence factor t, from $A_c(t,\tau)$ and denote it simply as $A_c(\tau)$

The mean DS of the channel, $\bar{\tau}$, is defined as,

$$\bar{\tau} = \frac{\int_{\tau=0}^{\infty} \tau A_c(\tau) d\tau}{\int_{\tau=0}^{\infty} A_c(\tau) d\tau} \tag{4.50}$$

The root-mean-squared (RMS) DS, τ_{rms}, is defined as,

$$\tau_{rms} = \sqrt{\frac{\int_{\tau=0}^{\infty} (\tau - \bar{\tau})^2 A_c(\tau) d\tau}{\int_{\tau=0}^{\infty} A_c(\tau) d\tau}} \tag{4.51}$$

The inverse of τ_{rms} is proportional to *coherence bandwidth*, which is a crucial parameter for waveform design.

τ_{rms} depends on several factors, first of which is obviously the dimension of the room, the larger is the dimensions, the more time the MPCs take to arrive at the Rx antenna, thus increasing the value of τ_{rms}, and also the reflection co-efficient of the reflecting surface determines the power of the reflected clusters thereby changing τ_{rms}. Also, increasing the gain of the Tx and Rx antennas would reduce τ_{rms}, by attenuating/blocking clusters which have a large AoA/AoD.

Power Angular Spread

mmW channels are mostly sparse in the angular domain due to diffused scattering, absorption, etc. Hence, multiple paths are grouped together in the spatial domain to form an angular cluster. The RMS angular spread signifies the angular distribution of the clusters and also the angular width of each cluster. The RMS angular spread is of two types, the overall angular spread, which denotes the AoA and AoD and the cluster angular spread which denotes the beamwidth corresponding to every AoD/AoA. The subpaths of a cluster all lie within the cluster angular spread.

The angular spread gives an indication of the spatial distribution of the signal. This in turn helps in designing of BF systems where the beamwidth can be adaptively controlled based on the angular spread for a particular user [HKL+13]. It is worth noting that the elevation angular spread is generally less than the azimuthal one for most applications and it generally decreases with distance. The cluster angular spread can be given as,

$$c_{ASi} = \sqrt{\overline{\theta^2} - (\bar{\theta})^2} \tag{4.52}$$

where,

$$\overline{\theta^2} = \frac{\sum_{k=1}^{M_i} \theta_k^2 P_k}{\sum_{k=1}^{M_i} P_k} \tag{4.53}$$

and,

$$\overline{\theta} = \frac{\sum_{k=1}^{M_i} \theta_k P_k}{\sum_{k=1}^{M_i} P_k} \tag{4.54}$$

Here, M_i is the number of subpaths for cluster i and θ_k and P_k are the angles and powers of the kth subpath.

Polarization

The most important term related to polarization is cross-polarized discrimination (XPD), which is defined as the ratio of the mean power received with co-polarized Tx and Rx antennas and power received with cross-polarized Tx and Rx antennas. A typical value of XPD for lower bands like 2.4 or 5 GHz is 10 dB in the LOS case and 3 dB in the NLOS case [MMSL10, E+04]. In mmW bands, there are already very few clusters and each cluster may have different polarization characteristics, so if the polarization characteristics of the Tx and Rx antennas are poorly matched, then a huge amount of performance degradation can occur.

In case of free space propagation, the polarization of the wave does not change, so co-polarized antennas should be used in Tx and Rx. In mmW bands, typical values of XPD are around 15–20 dB [MMSL10]. Another unique property, that is important in mmW frequencies, is that when a circularly polarized wave is incident on a reflecting surface with small angles, the handedness of polarization of the reflected wave is reversed [MSM+95] (i.e., the left-hand circular polarized (LHCP) wave becomes right-hand circular polarized (RHCP), and vice versa). This results in a polarization mismatch at Rx. As a result for circularly polarized mmW waves, the number of MPCs is very low.

K-Factor for wireless channel

For a LOS wireless channel, the Rician K-Factor is defined as

$$K_{rician} = \frac{\text{Power of LOS ray}}{\text{Sum of powers of all NLOS clusters}} \tag{4.55}$$

The K-factor is an important parameter as it is required to accurately predict the performance of a wireless system in a LOS channel [MDCC17]. The power of the LOS ray is relatively easy to estimate using the Friis equation [Fri46], however, the estimation of the power of NLOS clusters is dependent on the channel properties. A typical wireless channel like the modified Saleh model [SJJS00] is given by

$$h(t, \tau, \phi_t, \theta_t, \phi_r, \theta_r) = \sqrt{P_0}e^{j\kappa}\delta(t) + \beta(0,0)e^{j\alpha(0,0)}\delta(t)$$

$$+ \sum_{l=1}^{L}\sum_{k=0}^{K} \beta(k,l)e^{j\alpha(k,l)}$$

$$\times \delta(t - T_l - \tau_{k,l})\delta(\phi_r - \Phi_{r,l} - \phi_{r,k,l})$$

$$\times \delta(\phi_t - \Phi_{t,l} - \phi_{t,k,l})\delta(\theta_r - \Theta_{r,l} - \theta_{r,k,l})$$

$$\times \delta(\theta_t - \Theta_{t,l} - \theta_{t,k,l}), \tag{4.56}$$

In the above equation, P_0 is the power of the LOS ray, $\beta(k,l)$ is the magnitude of the kth ray of the lth cluster, T_l is the ToA of the lth cluster, and $\tau_{k,l}$ is the time difference between the arrival of the kth and zeroth ray of the lth cluster. Φ_l and Θ_l denote the azimuth and elevation angles of the lth cluster, $\phi_{k,l}$ and $\theta_{k,l}$ denote the azimuth and elevation angles of the kth ray of the lth cluster, and suffixes $_t$ and $_r$ denote AoD and AoA, respectively. κ and α are random variables uniformly distributed between 0 and 2π. As discussed earlier the ToA, AoAs, and AoDs, are random variables and their distributions depend on the specific channel model. For example, if we use the Saleh–Valenzuela model, T_l is Gamma distributed [MDCC17].

For most channel models, the average powers of rays are exponentially decaying as,

$$\overline{\beta^2(k,l)} = \overline{\beta^2(0,0)}e^{-\frac{\tau_{k,l}}{\gamma}} e^{-\frac{T_l}{\Gamma}}, \tag{4.57}$$

From (4.56) and (4.131), the mean total transmitted power can be calculated as

$$P_{total} = P_0 G_T(0,0)G_R(\phi_0, \theta_0) + \overline{\beta^2(0,0)}G_T(0,0)G_R(\phi_0, \theta_0)$$

$$+ \overline{\beta^2(0,0)}(\sum_{l=1}^{L}\sum_{k=0}^{K}\mathbb{E}[G_R(\phi_{r,k,l} - \phi_0, \theta_{r,k,l} - \theta_0)]$$

$$\times \mathbb{E}_{\{\theta_{r,k,l},\phi_{r,k,l}|\Theta_{r,l},\Phi_{r,l}\}}[G_T(\phi_{t,k,l}, \theta_{t,k,l})]\mathbb{E}_{\tau_{k,l}}[e^{-\frac{\tau_{k,l}}{\gamma}}]\mathbb{E}_{T_l}[e^{-\frac{T_l}{\Gamma}}])$$

$$= P_{LOS} + P_{NLOS}, \tag{4.58}$$

where G_T and G_R are the radiation patterns of the Tx and Rx antennas, respectively, and P_{LOS} and P_{NLOS} are the power of the LOS rays and the total power of NLOS clusters, respectively.

4.1.3 MIMO Spatial Channel Model

Introduction

As we have already seen, in the mmW domain, communication range is limited due to higher PL, so directional transmission is necessary. For that, it makes more sense to consider also the spatial domain along with the temporal. The first attempts to give a double-directional spatiotemporal channel model were made in [SMB01] and [MAH+06], where directionality was considered at both the Tx and the Rx and the CIR had both time delay as well as AoD and AoA elements. The CIR considering all the clusters is given as [KKA+14] for the lth cluster. It is seen here that the CIR accounts for the spatial departure and arrival angles in the terms Φ_l and Ω_l. Each path in the multipath clustered channel has an independent phase which depends on the wave propagation vector $\overrightarrow{\mathbf{k}_{lr}}$ and the receiver position $\overrightarrow{\mathbf{r}}$. When we consider all the clusters together, we obtain the resultant impulse response,

$$h(t, \tau, \Omega, \Phi) = \sum_{l=1}^{N_C} \alpha_l \delta(\tau - \tau_l) \delta(\Omega - \Omega_l) \delta(\Phi - \Phi_l) exp(j\overrightarrow{\mathbf{k}_l} . \overrightarrow{\mathbf{r}}) \quad (4.59)$$

where N_C is the number of clusters. Here $\overrightarrow{\mathbf{k}_l}$ is the wave propagation vector and $\overrightarrow{\mathbf{r}}$ is the antenna position vector. The channel gain coefficient α_l can be obtained for both single-, dual- and cross-polarized antennas. For polarized antennas, the CIR is a 2×2 matrix given as,

$$h(t, \tau, \Omega, \Phi) = \begin{bmatrix} h^{VV}(t, \tau, \Omega, \Phi) & h^{VH}(t, \tau, \Omega, \Phi) \\ h^{HV}(t, \tau, \Omega, \Phi) & h^{HH}(t, \tau, \Omega, \Phi) \end{bmatrix} \quad (4.60)$$

Here, the matrix elements denote the channel impulse between a particularly polarized antenna at the receiver and a particularly polarized antenna at the transmitter. For example, $h^{VH}(t, \tau, \Omega, \Phi)$ is the impulse response between the vertically polarized antenna at the receiver and the horizontally polarized antenna at the transmitter.

When considering the MIMO channel, the radio channel between the transmitter and receiver set of antennas is taken into account which is obtained by integrating over the spatial domains at both transmitter and receiver ends [KKA+14],

$$\mathbf{H}(t, \tau) = \int \overrightarrow{g_r}(\Phi)^T h(t, \tau, \Omega, \Phi) \overrightarrow{g_t}(\Omega) \overrightarrow{a_R}(\Phi)(\overrightarrow{a_T}(\Omega))^T d\Omega d\Phi \quad (4.61)$$

In the discrete domain, when the CIR from (4.59) is put in the above equation, the MIMO channel becomes sampled in the spatial and temporal domains and it becomes,

$$\mathbf{H}(t,\tau) = \sum_{l=1}^{N_C} \delta(\tau - \tau_l)\overrightarrow{g_r}(\Phi)^T h(t,\tau,\Omega,\Phi)\overrightarrow{g_t}(\Omega)\overrightarrow{a_R}(\Phi)(\overrightarrow{a_T}(\Omega))^T \quad (4.62)$$

Equation (4.62) is a preliminary result for spatial channel modeling which takes into account both 2D and 3D models. For 3D channels, the angles Φ and Ω are solid angles considering both azimuth and elevation angles. $\overrightarrow{g_r}(\Phi)$ and $\overrightarrow{g_t}(\Omega)$ are the antenna gains for the receiver and transmitter, respectively. $\overrightarrow{a_R}(\Phi)$ and $\overrightarrow{a_T}(\Omega)$ are the antenna array responses given by,

$$[\overrightarrow{a}_T(\Omega)]_i = exp(j\overrightarrow{\mathbf{k}}_{T,\Omega}.\overrightarrow{\mathbf{x}}_{t,i}) \quad (4.63)$$

$$[\overrightarrow{a}_R(\Phi)]_i = exp(j\overrightarrow{\mathbf{k}}_{R,\Phi}.\overrightarrow{\mathbf{x}}_{r,i}) \quad (4.64)$$

where $\overrightarrow{\mathbf{k}}_{T,\Omega}$ is the plane-wave vector on the corresponding antenna and $\overrightarrow{\mathbf{x}}_{t,i}$ is the position of the antenna. The array responses come from the phased array theory [PZ02] and give an indication of the phase difference across different antennas in the far field on the assumption.

For describing the complete spatial channel, many parameters are required. Each of these parameters has a statistical distribution from which their behavior can be known. Considering all of these parameters, the behavior of the entire channel can be obtained and important observations can be made. The parameters are defined as follows,

- **PDP:** The RMS DS, discussed in Section 4.1.2 depends strictly on the propagation conditions (LOS/NLOS) and also on the channel frequency. We show in Figure 4.6 the variation of RMS delay spread CDF with frequency for both scenarios.

  We see that RMS delay spread is significantly lesser for LOS conditions than NLOS and also there is a difference when we change the frequency from 28 to 73GHz. So, a system designed for a particular frequency may not be suitable for another mmW frequency and thus specific design parameters arise for every frequency.

- **RMS AoD/AoA spread:** We also plot the RMS angular spread CDF for both arrival and departure angles in Figures 4.7 and 4.8 and see that there is variation with respect to environment and frequency.

- **Cluster power and subpath powers:** The received cluster power depends on the path length of the cluster which in turn depends on the geometry of the clusters. The propagation effect is broadly divided into

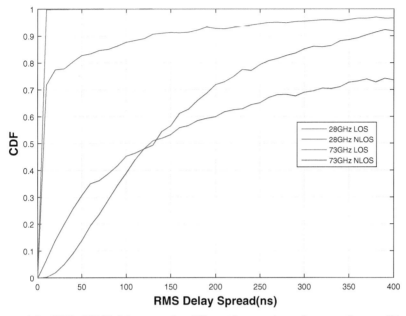

Figure 4.6 CDF of RMS delay spread at different frequencies and propagation conditions.

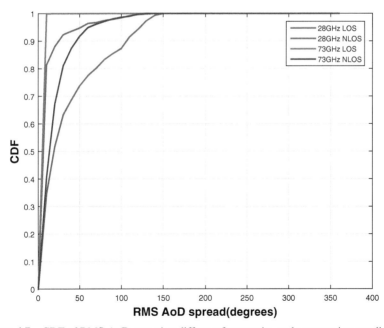

Figure 4.7 CDF of RMS AoD spread at different frequencies and propagation conditions.

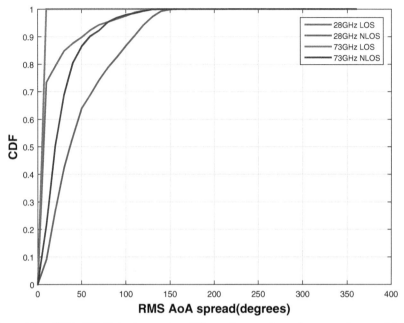

Figure 4.8   CDF of RMS AoA spread at different frequencies and propagation conditions.

two classes LOS and NLOS. The received powers for these two classes are vastly different and depend on the PL model [SNS$^+$14]. The received power is generally omnidirectional, and the cluster powers are obtained from it by using its envelop exponential decay. The cluster powers also suffer from large-scale variation due to a phenomenon called shadowing and generally follow a log-normal distribution [PP02].

- **Doppler Shift:** The Doppler shift denotes the change in the frequency of the signal received with respect to movement [Gol05]. It depends on the velocity and also the direction of movement with respect to the incoming signal direction. The Doppler frequency is different for different clusters as different clusters have different arrival angles. Doppler effect plays an important part in the design of physical layer as the Doppler shift affects the power spectral density (PSD) of the signals and spreads it in the frequency domain. Also, Doppler shift gives rise to an important parameter called coherence time [Gol05] which denotes how fast the signal undergoes small-scale fading and it directly gives an idea about the transmission strategy and symbol lengths to be used.

3GPP-based 3D channel model considering antenna polarization

Antenna polarization is a new diversity technique which is currently being considered for mmW communications. Polarization of signals changes when they encounter reflection or scattering. For example, a vertically polarized signal may not remain vertically polarized after reflection. This aspect is brought forward by bringing the interference caused by polarization mismatch in the 3D model. When vertical and horizontal polarizations are used together in the same antenna (cross-polarization), it provides diversity. But in the receiver, the signals are not perfectly horizontally or vertically polarized. Thus, a cross-polarized antenna at the receiver will receive interference from the polarization direction perpendicular to it, i.e., the vertically polarized part will receive the transmitted vertically polarized signal as well as some part of the transmitted horizontally polarized signal with a phase difference. The channel model considering polarization had been adopted in WINNER II and 3GPP [ATR16a] and is proposed for mmW in [TNMR14b], already discussed in Section 4.1.2.

A cross-polarized mmW channel can be given as

$$
\mathbf{H}_{u,s,n}(t) = \sqrt{P_n} \sum_{m=1}^{M} \begin{bmatrix} F_{rx,u,V}(\psi_{n,m}, \Psi_{n,m}) \\ F_{rx,u,H}(\psi_{n,m}, \Psi_{n,m}) \end{bmatrix}^T
$$

$$
\times \begin{bmatrix} e^{j\phi_{n,m}^{vv}} & \sqrt{\kappa_{n,m}^{-1}} e^{j\phi_{n,m}^{vh}} \\ \sqrt{\kappa_{n,m}^{-1}} e^{j\phi_{n,m}^{hv}} & e^{j\phi_{n,m}^{hh}} \end{bmatrix} \times \begin{bmatrix} F_{tx,s,V}(\theta_{n,m}, \Theta_{n,m}) \times \\ F_{tx,s,H}(\theta_{n,m}, \Theta_{n,m}) \end{bmatrix}
$$

$$
\times e^{j2\pi\lambda_0^{-1}(\overline{\phi}_{n,m}.\overline{r}_{rx,u})} e^{j2\pi\lambda_0^{-1}(\overline{\theta}_{n,m}.\overline{r}_{tx,v})} e^{j2\pi\nu_{n,m}t} \tag{4.65}
$$

Here the term $F_{tx/rx,antenna,direction}$ represents the antenna gain for the particular TX/RX antenna in the vertical or horizontal direction obtained from the radiation pattern at the particular elevation and arrival angles. The matrix in the middle is the polarization matrix which represents the cross-polarization interference terms. As a polarized wave hits a scatterer, it undergoes a phase change. The terms $\phi_{n,m}^{x,y}$ denote the phase change as the polarization changes from y to x. The term $\kappa_{n,m}$ is the cross-polarization power ratio which indicates the ratio between the two polarization components. If it is not zero, then it indicates that there is cross-polarization distortion present. When the signal paths hit the antenna, they cause a phase difference across the antenna elements. This phase difference can be modeled considering narrowband antennas in the far field. The components, $\overline{\phi}_{n,m}.\overline{r}_{rx,u}$ and $\overline{\theta}_{n,m}.\overline{r}_{tx,v}$ denote the antenna phases for the receiver and transmitter, respectively, where $\overline{\phi}_{n,m}$ and $\overline{\theta}_{n,m}$ are the unit vectors in the direction of the receiver and transmitter signal paths, given as,

$$\overline{\phi}_{n,m} = \begin{bmatrix} sin(\Psi_{n,m})cos(\psi_{n,m}) \\ sin(\Psi_{n,m})sin(\psi_{n,m}) \\ cos(\Psi_{n,m}) \end{bmatrix} \quad (4.66)$$

while $\overline{r}_{rx/tx,u/v}$ is the vector of the position of the TX/RX antenna. The last term in 4.145 is the phase shift due to Doppler which occurs when there is movement. It is highly dependent on time and also the direction of the arriving signal path at the receiver with respect to direction of movement given by the velocity vector. The Doppler frequency is given as,

$$\nu_{n,m} = \frac{\overline{\phi}_{n,m}^{T}\overline{\nu}}{\lambda_0}$$

$$\overline{\nu} = \nu_s \begin{bmatrix} sin(\theta_\nu)cos(\phi_\nu) & sin(\theta_\nu)sin(\phi_\nu) & cos(\theta_\nu) \end{bmatrix}^{T} \quad (4.67)$$

where θ_ν and ϕ_ν are the elevation and azimuthal angles of direction of movement with respect to the incoming signal. A typical scattering environment involving a particular cluster and uniform planar arrays has been shown in Figure 4.16 to give an understanding of the geometry involved.

As per the 3GPP model, a uniform planar antenna array is composed of antenna ports in the horizontal direction and antenna elements in the vertical direction. Each antenna port carries the signal to be fed into different antenna elements. For obtaining the net channel matrix, a complex weight is applied in the vertical domain on each element given as [ATR16b],

$$\omega_q = \frac{1}{\sqrt{Q}}exp\left(-j\frac{2\pi}{\lambda}(q-1)d_\nu cos(\theta_{tilt}) \right) \quad (4.68)$$

where Q is the number of antenna elements per port. θ_{tilt} is the tilt angle with respect to the vertical axis. The overall channel considering BF at both the transmitter and the receiver is given as,

$$[H_{i,n}^c(t)]_{a,b} = \sum_{u \in P_a} \omega_s \sum_{s \in P_b} \omega_u H_{i,u,s,n}(t) \quad (4.69)$$

for receiver antenna port a and transmit antenna port b. P_a and P_b are the sets of antenna elements corresponding to ports a and b, where $a \in \{1, 2, ...N_{RX}\}$ and $b \in \{1, 2, ...N_{TX}\}$.

Virtual Channel representation for MIMO channels

The physical modeling of the spatial MIMO channel gives an idea about the positions of clusters. The MIMO channel in mmW domain is sparse and so, for practical signal processing techniques, the complexity can be reduced by considering lower dimensional channel matrices using beam selection and combining. For this reason, an alternate representation has been developed. This alternate representation is called virtual channel representation [Say02] and it helps in finding out the spatial degrees of freedom in a MIMO channel which in turn helps in calculation of capacity and level of diversity gain. It gives an idea about the interaction between the clusters by showing how many transmitter cluster paths are associated with a receiver path and vice versa. It also shows the number of spatial degrees of freedom of the channel and, hence, the number of distinct orthogonal beams which can be formed. The virtual representation expresses the channel matrix elements as a Fourier transform in the spatial domain of the virtual channel matrix elements. In other words, the entire channel is represented as a combination of orthogonal spatial basis functions with each coefficient in the beamspace [BBS13] domain. The representation of the channel is given in [Say02] as,

$$H(m, n) = \sum_{q=-Q}^{Q} \sum_{p=-P}^{P} H_v(p, q) e^{-j2\pi mq/Q} e^{j2\pi np/P} \qquad (4.70)$$

where $H_v(p, q)$ is the channel coefficient in the spatial beamspace domain and P and Q are the number of spatial signal vectors at the transmitter and the receiver. The virtual representation gives an insightful imaging interpretation of the scattering geometry. A realistic channel which has multiple scatterers, each with limited cluster angle spread, can be represented by the virtual matrix consisting of non-vanishing sub-matrices corresponding to different clusters. The dimensions of the sub-matrices denote the extent of the angular spread of each cluster and the nature of each sub-matrix gives an idea about the number of distinct, independent paths in each cluster. The number of sub-matrices effectively gives the number of spatial degrees of freedom and the number of effective beams that can be formed which help in determining the capacity and also selecting the appropriate BF vectors [BBS13, GDC+16, AM15].

When representing the channel in the virtual domain for a linear antenna array, the transmitter and receiver angular domains are divided uniformly, i.e., the transmitted paths have P uniformly spaced angles within $[\frac{-\pi}{2}, \frac{\pi}{2}]$ and the receiver is divided into Q uniformly spaced angles. Then the linear array vectors corresponding to all the transmitter and receiver angles form two sets of orthogonal basis. The basis vectors are of the form,

$$\mathbf{a}(\theta_T) = \frac{1}{\sqrt{P}}[1, e^{-j2\pi\theta_T}, e^{-j4\pi\theta_T}, \ldots\ldots\ldots, e^{-j2\pi(P-1)\theta_T}] \qquad (4.71)$$

$$\mathbf{a}(\theta_R) = \frac{1}{\sqrt{Q}}[1, e^{-j2\pi\theta_R}, e^{-j4\pi\theta_R}, \ldots\ldots\ldots, e^{-j2\pi(Q-1)\theta_R}] \qquad (4.72)$$

The angles of the basis vectors are uniformly distributed, i.e., $\theta_T = \frac{p}{P}$ where $p \in [-P, P]$ and $\theta_R = \frac{q}{Q}$ where $q \in [-Q, Q]$. If the radio channel is expressed as,

$$\mathbf{H} = \int_{-\alpha_R}^{\alpha_R} \int_{-\alpha_T}^{\alpha_T} G(\theta_R, \theta_T)\mathbf{a_R}(\theta_R)\mathbf{a_T}(\theta_T)d\theta_R\theta_T \qquad (4.73)$$

where $G()$ is the channel gain at a particular pair of transmit and receive path angles which takes into account the gains of the antennas also and α_R and α_T denote the angular spread at the receiver and the transmitter.

The beamspace domain channel coefficients are then represented as,

$$H_v(q, p) \approx \frac{G(\frac{q}{Q}, \frac{p}{P})}{PQ} \qquad (4.74)$$

Since the transmitter/receiver domains are sampled uniformly, the integration becomes summation in the discrete domain. Thus, the channel is represented as,

$$\mathbf{H} = \sum_{q=-Q}^{Q} \sum_{p=-P}^{P} H_v(q, p)\mathbf{a}(\theta_{R,q})\mathbf{a^H}(\theta_{T,p}) \qquad (4.75)$$

$H_v(q, p)$ is the channel gain corresponding to the spatial bin defined by $\frac{q}{Q}$ and $\frac{p}{P}$ based on the following conditions,

$$S_{R,q} = l : \frac{-1}{2Q} \leq \theta_{R,l} mod1 - \frac{q}{Q} \leq \frac{1}{2Q} \qquad (4.76)$$

$$S_{T,p} = l : \frac{-1}{2P} \leq \theta_{T,l} mod1 - \frac{p}{P} \leq \frac{1}{2P} \qquad (4.77)$$

Here, $S_{R,q}$ and $S_{T,p}$ are the sets formed at the receiver and the transmitter respectively, by grouping of cluster paths within $\frac{1}{Q}$ or $\frac{1}{P}$ angular spread about a particular virtual angle. The cluster paths associated with a particular transmitter–receiver pair of virtual angles, each has a gain of β_l which depends upon the PL, reflection loss, etc., are grouped together and the virtual channel coefficient is given by the sum of the cluster path powers.

$$H_v(q, p) \approx [\sum_{l \in S_{T,p} \cap S_{R,q}} \beta_l] \tag{4.78}$$

From the virtual channel representation, an approximate measure of the order of the channel and the diversity can be obtained. The order of the channel or the number of parallel sub-channels is given by the number of non-zero components in the sub-matrices while the diversity is given by the number of virtual transmit angles associated with each virtual receive angle and vice versa.

The virtual representation also gives a way for BF in lower dimensions. Many BF methods like in [BBS13] and [HZJ$^+$] represent the channel in the virtual domain in order to directly select the BF vectors from the spatial basis. We note that the beamspace channel can be easily represented as,

$$\mathbf{H} = \mathbf{U}_R \mathbf{H}_v \mathbf{U}_T{}^H \tag{4.79}$$

where $\mathbf{U}_R$ and $\mathbf{U}_T$ are the basis matrices in the beamspace domain for the receiver and transmitter, respectively. The columns of the matrices are simply the array vectors given in (4.71) and (4.72) at all the virtual angles if we consider uniform linear arrays. These matrices are analogous to DFT matrices, and thus the physical channel matrix can be simply obtained from the beamspace one by simple Fourier transform. Thus, when considering the linear model for the MIMO system like,

$$\mathbf{Y} = \mathbf{H}\mathbf{X} + \mathbf{N} \tag{4.80}$$

it can be represented in the beamspace domain as,

$$\mathbf{Y}_v = \mathbf{H}_v \mathbf{X}_v + \mathbf{N}_v \tag{4.81}$$

where $\mathbf{Y}_v = \mathbf{U}_R{}^H \mathbf{Y}$ and $\mathbf{X}_v = \mathbf{U}_T \mathbf{X}$ are the spatial domain representations of the output and input vectors. By, this model, analog as well as hybrid BF [EARAS$^+$14] methods can be applied to maximize the SNR for a particular user or a group of users.

For 5G mmW MIMO systems, the antenna arrays used are mainly 2-D, and so the beamspace should consider both azimuthal and elevation domains.

If the number of azimuthal (horizontal) antennas is N_h and the number of vertical antennas is N_v, then the beamspace is built by the following vectors [HZJ$^+$, BS14],

$$\mathbf{a}^h(\phi_i) = \frac{1}{\sqrt{N_h}}[1, e^{-j2\pi\phi_i}, \ldots\ldots\ldots, e^{-j2\pi(N_h-1)\phi_i}]^T \tag{4.82}$$

$$\mathbf{a}^v(\theta_i) = \frac{1}{\sqrt{N_v}}[1, e^{-j2\pi\theta_i}, \ldots\ldots\ldots, e^{-j2\pi(N_v-1)\theta_i}]^T \tag{4.83}$$

Thus, the beamspace conversion matrices U_T and the U_R at the transmitter and the receiver are obtained by the following basis vectors,

$$\mathbf{a}_{T/R}(\phi_i, \theta_i) = \mathbf{a}^h(\phi_i) \otimes \mathbf{a}^v(\theta_i) \tag{4.84}$$

The matrices $\mathbf{U}_T$ and $\mathbf{U}_R$ are obtained by the same way as for 2-D, by sampling both the azimuthal and elevation domains uniformly. Hence,

$$U_{T/R}(p,q) = \mathbf{a}_{T/R}\left(\frac{p}{N_h}, \frac{q}{N_v}\right)$$
$$p \in \{0, 1, \ldots, N_h - 1\}, q \in \{0, 1, \ldots, N_v - 1\} \tag{4.85}$$

To give an actual understanding of the beamspace domain, we plot the beamspace channel matrix for three users separated by a certain angle on a plane with a BS having a 8×8 UPA in Figure 4.9. We consider the uplink channel and see that the beamspace channels have certain regions where the energy is significantly higher than the surroundings. These areas are significantly small as compared to the total spatial domain. So, with appropriate selection of the virtual angles, we can reduce the dimension of the channel considered by a significant amount. The selected virtual angles are used to form the BF matrix. However, the identification of the sparse nature of the channel in the spatial domain is a challenging task and gives rise to antenna selection algorithms [AM15].

Beamspace-based MIMO communication can exploit the sparsity of mmW channels for reducing the amount of data required for transmitting channel information, number of channel estimation pilots and beam-selection matrices. This helps in low complexity implementation of multiplexing protocols like spatial division multiple access (SDMA). In [HZJ$^+$], an enhanced SDMA scheme has been proposed, which makes use of beamspace domain channel modeling to reduce the overhead of channel information acquisition and interference suppression precoding. Authors in [DS14] have used the

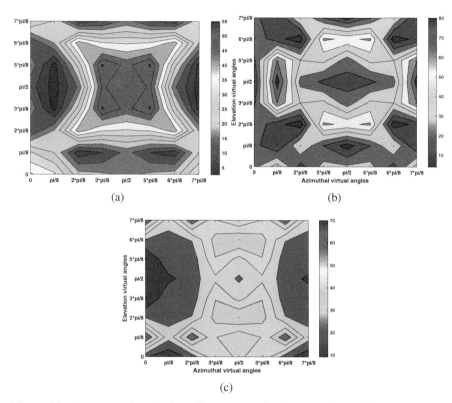

Figure 4.9 Beamspace domain channel energy map for three users in a uniform 8×8 planar array BS scenario.

beamspace model for reducing the complexity of mobile station localization in a cell. A maximum-likelihood classification method has been developed with near-optimal performance and very low complexity. In another work in [BS15], a differential transmission scheme has been used along with beamspace MIMO which allows working with quasi-coherent channel estimates in order to reduce the complexity of beam selection algorithms. This differential beamspace MIMO enables low-complexity transceiver implementation for mmW using only differential channel estimates with minimum performance loss. The work has been extended in [SB15], where channel models for the differential MIMO have been developed and interference suppression techniques have been stated.

Another advantage of using this domain especially in the MU case is that the precoder complexity can be greatly reduced without affecting the sum capacity in the downlink by a big amount. The work in [BS14] shows this where there is negligible capacity loss with a significant reduction in complexity.

Thus, we see that signal processing for mmW MIMO systems can be done with very low complex methods utilizing the beamspace domain.

Massive MIMO Spatial Channel Modeling

In massive or large-scale MU MIMO, a BS with several hundred antennas (assumed to be $= M$) serves several tens (assumed to be $= K$) UEs (the UEs are sufficiently apart, so the channels observed by any two UEs are uncorrelated) simultaneously. We assume throughout this section that $M >> K$. The advantages include [ZOY14, LETM13, RPL$^+$13] high SE, high energy efficiency, high reliability, efficient encoder design, low interuser interference, etc.

The massive MIMO channel modeling is divided into correlation-based stochastic models (CBSMs) and geometry-based stochastic models (GBSMs).

CBSM

Just like in the SISO case, CBSM assumes probability distributions for channel modeling. So, it produces models which are analytically tractable. If we assume that K UEs have one antenna each, then we can model the reverse link (RL) channel as [RPL$^+$13, ZOY14]

$$\mathbf{G} = \mathbf{H} \mathbf{D}_\beta^{1/2} \qquad (4.86)$$

where $\mathbf{D}_\beta$ is the large-scale propagation matrix which defines the large-scale fading effects and is a diagonal matrix defined as

$$\mathbf{D}_\beta = diag\{\beta_1, \beta_2, ..., \beta_K\} \qquad (4.87)$$

where $\beta_k = \phi d_k^{-\alpha} \xi_k$, ϕ is a constant related to antenna gain, d_k is the distance between the BS and the kth UE, α is the PLE, and ξ_k is the large scale log-normal distributed shadow fading. $\mathbf{H}$, on the other hand, is an $M \times K$ matrix which denotes the small scale fading and its coefficients are circularly Gaussian.

The system model of the RL in the massive MU-MIMO system can be written as,

$$x_r = \sqrt{\rho_r} G q_r + w_r \qquad (4.88)$$

where x_r is the transmitted signal in the RL and q_r contains the encoded data of the r user.

The capacity of the channel in RL is given by

$$C = log_2\ det(\mathbf{I}_K + \rho_r \mathbf{G}^H \mathbf{G}) \tag{4.89}$$

where ρ_r is the SNR of the individual channels.

Under favorable propagation conditions, when the number of antennas is increased, the capacity expression becomes a sum of K terms because,

$$(\frac{\mathbf{G}^H \mathbf{G}}{M})$$
$$= \mathbf{D}_\beta^{1/2}(\frac{\mathbf{H}^H \mathbf{H}}{M})\mathbf{D}_\beta^{1/2}$$
$$\approx \mathbf{D}_\beta \tag{4.90}$$

So, the capacity expression in (4.89) becomes,

$$C_{sum_r, M>>K} \approx log_2 det(\mathbf{I_K} + M\rho_r \mathbf{D}_\beta)$$
$$= \sum_{k=1}^{K} log_2(1 + M\rho_r \beta_r) \tag{4.91}$$

So, here we see that under a large number of antennas, the channel matrix approximately becomes uncorrelated which can help in obtaining full spatial diversity for serving K users. This is because, under the law of large numbers as $M \to \infty$, $\frac{\mathbf{H}^H \mathbf{H}}{M} \to \mathbf{I_K}$. The term *favorable propagation conditions* implies the Gaussian assumption on the distribution of the elements of $\mathbf{H}$.

Similarly, for the forward link (FL), we can modify (4.86) as

$$\mathbf{G} = \mathbf{HD}_\gamma^{1/2} \tag{4.92}$$

where

$$\mathbf{D_{fl}} = diag\{\gamma_1, \gamma_2, ..., \gamma_M\} \tag{4.93}$$

The sum capacity at FL is

$$C_{sum_f} = \underset{\gamma_k}{max}\ log_2\ det(\mathbf{I}_M + \rho_f \mathbf{GD}_\gamma \mathbf{G}^H)$$
$$subject\ to \sum_{k=1}^{K} \gamma_k = 1, \gamma_k \geq 0, \lor k \tag{4.94}$$

because, as per reciprocity, the FL channel is the transpose of the RL channel.

Under a very large number of antennas, FL capacity assuming favorable propagation conditions can be written as,

$$C_{sum_f} = \max_{\gamma_k} \ log_2 \ det(\mathbf{I}_M + \rho_f \mathbf{GD}_\gamma \mathbf{G}^H)$$

$$= \max_{\gamma_k} \ log_2 \ det(\mathbf{I}_M + \rho_f \mathbf{D}_\gamma^{1/2} \mathbf{G}^H \mathbf{GD}_\gamma^{1/2})$$

$$\approx \max_{\gamma_k} \ log_2 \ det(\mathbf{I}_M + \rho_f M \mathbf{D}_\gamma \mathbf{D}_\beta)$$

$$= \max_{\gamma_k} \ \sum_{k=1}^{K} log_2(1 + M\rho_f \gamma_k \beta_k) \tag{4.95}$$

This is an optimization problem which can be solved by allocating the powers γ_k for the users such that the total power constraint is maintained. But we see that under massive MIMO, both the RL and FL channels can be diagonalized and the capacity expression can be broken down to K terms. This shows that there is a Kth order spatial diversity and capacity increases linearly with increasing the number of UEs in both RL and FL. This is the most ideal situation for massive MIMOs, as apart from high capacity, it simplifies massive MIMO algorithm design and reduces interuser/intercell interference [Zoy14].

Correlation Channel Model

If the antenna spacing is small, then we cannot assume that the channels observed by two adjacent Tx antennas are uncorrelated. To model the correlation, the most commonly used geometry is the two-ring scattering model, which models two sets of independent scatterers around the transmitter and the receiver. This model is suitable for most macrocellular applications and is given as,

$$\mathbf{G} = \mathbf{\Psi_r}^{1/2} \mathbf{G}_{iid} \mathbf{\Psi_t}^{1/2} \tag{4.96}$$

The correlation matrices $\mathbf{\Psi_r}$ and $\mathbf{\Psi_t}$ are independent and $\mathbf{G}_{iid}$ is the normal Rayleigh faded channel matrix. The channel correlation matrix can be obtained as $\mathbf{\Psi} = \mathbf{\Psi_r} \otimes \mathbf{\Psi_t}$ and the channel matrix can be obtained as,

$$vec(\mathbf{G}) = \mathbf{\Psi} vec(\mathbf{G}_{iid}) \tag{4.97}$$

where $vec()$ operation stacks vertically the columns of a matrix. The Kroenecker-based model is easy to analyze because it separates the transmitter and receiver domains, but it does not model accurately the correlation between the Tx/Rx antennas and the joint correlation between the transmitter and the receiver which may occur in some scenarios. Hence, the Weich-selberger model has been defined which has the ability to include antenna correlations at both the transmitter and the receiver. Furthermore, it jointly considers the correlation between the transmitter array and the receiver array. The joint correlation is modeled by a coupling matrix and hence the channel model is written as,

$$\mathbf{H}_{model} = \mathbf{U}_A(\tilde{\boldsymbol{\Omega}} \odot \mathbf{G})\mathbf{U}_B^T \tag{4.98}$$

where $\mathbf{U}_A$ and $\mathbf{U}_B$ denote the transmitter and receiver spatial eigenbasis matrices having elements analogous to the phase coefficients of an n antenna array at the particular cluster angle. $\mathbf{G}$ is the normal Rayleigh fading channel matrix and $\tilde{\boldsymbol{\Omega}}$ is the coupling matrix which couples the power between the transmitter and receiver antennas. $\odot$ denotes elementwise multiplication of two matrices. When the coupling effect is not present, this model reduces to the Kroenecker model.

Mutual Coupling Channel Model

As the antenna array size increases while the aperture remains fixed, the antenna spacing decreases and this leads to electromagnetic coupling effects. These effects can be modeled by considering the MIMO system as two-port networks with antenna impedance matrices and matching networks. The channel vector for the kth user in the MU-MIMO system under both receiver correlation and coupling is defined as,

$$\mathbf{h}_k = \mathbf{Z}\mathbf{R}_k\mathbf{v}_k \ k = 1, 2, ...K \tag{4.99}$$

where $\mathbf{Z}$ is the coupling matrix, $\mathbf{R}_k$ is the antenna correlation matrix at the receiver, and $\mathbf{v}_k$ is a zero-mean unit variance Gaussian vector. The coupling matrix is given as,

$$\mathbf{Z} = (Z_A + Z_L)(\boldsymbol{\Gamma} + Z_L\mathbf{I})^{-1} \tag{4.100}$$

with

$$\mathbf{Z} = \begin{bmatrix} Z_A & Z_M & 0 & \cdots & 0 \\ Z_M & Z_A & Z_M & \cdots & 0 \\ 0 & Z_M & Z_A & \cdots & 0 \\ \vdots & \vdots & \ddots & \ddots & \vdots \\ 0 & 0 & \cdots & Z_M & Z_A \end{bmatrix} \tag{4.101}$$

where Z_A, Z_L, and Z_M represent the antenna impedance, load matching impedance, and the mutual coupling impedance. The mutual impedances are considered only between adjacent antennas.

GBSM

Geometry-based stochastic models use the propagation environment to estimate the spatial and temporal parameters like arrival times and AoD/AoA of clusters. Based on these data, a channel model is created through ray tracing. GBSMs are more useful for analyzing the performance of practical communication systems. They may not be analytically tractable like CBSMs, but still some conclusions can be made by the knowledge of the temporal and angular distributions. GBSMs can be in both 2-D and 3-D since 3-D model is needed for future array models like cylindrical and planar arrays. The models for normal arrays resemble those of the conventional standardized models like the 3GPP. But when considering massive MIMO, these models have to be modified because, in massive antenna arrays, the phase difference between the elements is not constant because the size of the array is bigger than Rayleigh distance and so, instead of plane wavefront, a spherical wavefront assumption is taken. In [WWA+14], a twin-cluster non-stationary MIMO channel model has been considered, which takes the spherical wavefront assumption and finds the dependence of different cluster parameters on the antenna correlations. In [PT12], channel measurements were done at 2.6 GHz and it was found that the Rayleigh distance was very high and so, non-stationarity was observed along the array. Additionally in [WWH+15], the non-stationary model was analyzed for elliptical distribution of clusters. The change in the channel parameters was modeled for movement of clusters which caused displacement of the ellipses.

Another effect which occurs in large antenna arrays apart from the spherical wavefront assumption is that the clusters tend to appear and disappear along the array axis, i.e., adjacent antennas can observe different sets of clusters. This generally results in a decrease of spatial correlation which may prove more beneficial for capacity. The appearance and disappearance of clusters are usually modeled as a birth-death process along the spatial dimensions [WWH+15]. According to that, the probability of survival of a cluster when going from one antenna to other is given as,

$$P_{survival}^{T/R} = e^{-\lambda_R \frac{\delta_{T/R}}{D_c^a}} \tag{4.102}$$

where $\delta_{T/R}$ is the antenna spacing at the transmitter/receiver and D_c^a is the scenario-dependent correlation factor. As a result of this modeling, the cluster sets evolve from antenna to antenna. New clusters are generated and existing clusters in the set may survive or annihilate never to return again.

The expected number of new clusters generated is,

$$E[N_{new}^{T/R}] = \frac{\lambda_G}{\lambda_R}(1 - e^{\frac{-\delta_{T/R}}{D_c^a}})$$ (4.103)

where λ_G is the cluster generation and λ_R is the cluster recombination rate. The cluster sets are obtained for every antenna one by one for both transmitter and receiver, and then for every cluster, the transmitter and receiver antennas which observe that cluster are grouped together. The extra clusters which are not observable by both the transmitter and the receiver are eliminated. The detailed process has been described in [WWH+15].

A direct effect of modeling this cluster evolution process along the array happens on the correlation between two antenna links. If we consider two pairs of antennas k, l and k', l' with $k, k' \in transmitter$ and $l, l' \in receiver$, then the correlation coefficient between the two links under this cluster generation process can be written as,

$$\rho_{kl,k'l',n}^{LOS/NLOS}(\delta_T, \delta_R; t) = e^{\lambda_R \frac{-|l-l'|\delta_T - |k-k'|\delta_R}{D_c^a}} \rho_{kl,k'l',n}^{LOS/NLOS1}(t)$$ (4.104)

where $\rho_{kl,k'l',n}^{LOS/NLOS1}(t)$ is the correlation without assuming the non-stationary effects.

4.2 Millimeter-Wave Communications

The history of mmW communications goes back to 1890s when Sir J. C. Bose experimented with it [Bos27]. For the next 50 years, mmW system research was limited mostly in academics and research labs. Later mmWs started to be used first in radio astronomy in 1960s, then followed by the military in 1970s. In 1980s with the advent of microwave monolithic integrated circuits (MMICs), mmW gradually came into commercial use [Adh08]. One of the first commercial applications of mmW was automotive collision avoidance radar. Federal Communications Commission (FCC) opened up an industrial, scientific, and medical (ISM) band between 59 and 64 GHz in 1995. This band includes the well-known 60-GHz band, which is considered ideal for short distance communication due to the high attenuation caused by atmospheric oxygen molecules [RMG11]. On the other hand, very low atmospheric attenuation between 92 and 95 GHz has made this band very popular for mmW radar [HH69]. More recently, in 2004, FCC in an order [FCC04] has allocated 5 GHz between 71 and 76 GHz for downlink and an equal amount of bandwidth between 81 and 86 GHz for uplink in fixed satellite services (FSS). Lower atmospheric attenuation has made these bands very attractive even for geostationary orbit satellites [SMC$^+$12].

All the previous mobile standards from first-generation (1G) to 4G have used sub-6-GHz bands. On the other hand, a lot of relatively unused bandwidth is present in microwave and mmW frequencies between 3 and 300 GHz. Some of the popular frequency bands are 28–30 GHz where extensive modeling through measurements of both indoor [MRSD15] and outdoor [NMSR14, Z$^+$13, AWW$^+$13, RSM$^+$13b] wireless channels, the license-free frequency of 60 GHz [YXVG11], where IEEE 802.11ad WLAN standard has been already finalized [agi13], and future wireless LAN standards like IEEE 802.11aj, IEEE 802.11ay, and IEEE 802.11az will also use mmW. FCC has also allocated two bands of 71–76 and 81–86 GHz for FSS [SMC$^+$12], out of which 73 GHz has found interest from 5G researchers [MRSD15, NMSR14, Z$^+$13, AWW$^+$13, RSM$^+$13b]. Research works [SMC$^+$12] have shown high throughputs even for geostationary satellite links at mmW. Even in terrestrial mobile communication systems, several works like [NMSR14, MRSD15] and [RSM$^+$13b] show appreciable signal strengths even in most NLOS links for both indoor and outdoor links.

Space division multiplexing by using multiple antennas at the transmitter (Tx) and the receiver (Rx) [PNG03a] is rapidly gaining popularity to design high communication systems with very high spectrum efficiency systems. For MU MIMO systems, it is known that the sum rate capacity of the Gaussian broadcast channel is limited by N_T [GKH$^+$07], which is the number of antennas in the BS. This gives an unique advantage in mmW range, as smaller antennas help us to pack more antennas in the BS giving us higher capacity. Also, a large number of antennas in the BS would help to create sharp beams using analog, digital, and hybrid BF [HGPR$^+$16]; this increased antenna gain will compensate the slightly higher path loss at mmW range coming from the Friis equation [Fri46]. It is known that the rms delay spread [LSE07, MDCC17] for mmW frequencies is much smaller than that of sub-6-GHz bands. This means that the coherence bandwidth for mmW channels is much larger than sub-6-GHz channels. This will enable us to design OFDM systems with larger subcarrier bandwidth, this would keep the distortions due to Doppler shift (which is proportional to center frequencies) low [YXVG11].

4.2.1 Challenges

The use of mmW in cellular scenarios specially when a direct LOS component is not available is still a matter of research. The wider bandwidths of mmW does offer higher throughputs however, mmW signals usually suffer larger shadowing, and higher Doppler spreads [RRE14, RSM+13a, DMRH10]. However, the advent of low-cost mmW components using CMOS has made design easier and more cost effective [RMG11]. In addition, small wavelengths in mmW enable larger arrays to be fabricated in smaller areas. Human blockage is another big problem in mmW and MAC algorithms are being developed to solve it [RRE14]. Also, as the data rates increase, the cell sizes tend to go smaller [RRE14, MMW+15, ACD+12, CD14] with the advent of picocells and femtocells. So it can be expected that soon the radii of the cells would be in the order of 200 m which is within the range of most microwave measurements [RSM+13a]. In the absence of spectrum, attempts to increase the capacity of networks will require even higher densification of cells. Extreme cell densification without looking for newer frequencies may not be cost effective in many settings due to due to higher capital expenditure and BH requirements. On the other hand, a combination of network densification and mmW may remarkably enhance the capacity of individual small cells. Higher frequencies is an interesting option even for BH [RRE14].

The major challenges of mmW communications are

- Range: As we have seen earlier the FSPL increases with frequency. Fortunately, so does the antenna gain for the same antenna size.
- Shadowing: mmWs are highly susceptible to shadowing and blocking by human beings and other objects. Brick walls can attenuate mmW signals by 40–80 dB [Z+13, ADH+94] and human body can attenuate mmW signals between 20 and 35 dB [LSCP12]. On the other hand, rain fades and atmospheric absorptions are also major problem at this frequency [GLR99, QL06]. On the other hand, strong NLOS links exist, in sub-6-GHz frequencies, due to the presence of a very large number of multipaths and lower gain antennas, so blockage is less a problem.
- Rapid channel fluctuations and intermittent connectivity: For a given mobile velocity, channel coherence time is linear in the carrier frequency [Rap02]. For example, Doppler spread at a velocity of 60 km/h at 60 GHz would be around 3.3 kHz. Since coherence time is linked with

Doppler spread by the following formula [Rap02]

$$T_c \simeq \frac{0.423}{f_d} \qquad (4.105)$$

where T_c is the coherence time and f_d is the Doppler spread, so a Doppler spread of 3.3 kHz will give a coherence time of around 120 μs, which is much faster than today's cellular systems. Also, high amount of shadowing due to objects will mean that the amount of received power would change much more dramatically than in conventional sub-6-GHz systems. This means that communication should be highly intermittent and our system should be highly adaptive [RRE14].

- MU Coordination: Although mmW is currently limited to point to point links for higher SE applications frequency reuse has to be used and hence new mechanisms to enhance the signal to interference ratio have to be encorporated.

- Processing Power Consumption: The high bandwidth available at mmW frequencies will require very high speed sampling at analog to digital converters (ADCs) of several gigasamples per second (Gsps), which will lead to very high power consumption. High bandwidth requires higher sampling rates as per the Nyquist sampling theorem [RMG11, CCCG94, MR14], this leads to high power consumption and/or highly expensive ADCs. This makes incorporation of a large number of antennas and multibit quantization difficult for cheap handheld devices.

- Device Non-idealities: Various device non-idealities like phase noise in amplifiers (PA) and non-linearities in power amplifier (PA) increase with frequency. These have a severe effect on the performance of communication systems [SCM+13, DCR+15]. This has to be mitigated by using smarter design techniques.

Most practical applications of mmW has used small cell and/or outdoor stationings, such as campuses, stadiums, etc [PBR+12]. Propagation issues discussed earlier make it difficult for mmW systems to provide fast and reliable communications. It is predicted [RRE14] that mmW systems would be heterogeneous. Interference management in Heterogeneous networks, or HetNets, have been one of the most sought research areas in recent years [CAG09, ACD+12]. Some of the interesting consequences of introduction of mmW would be [RRE14].

- Co-existence with sub 6 GHz systems: Good coverage would require both mmW and conventional sub 6 GHz cellular infrastructure.
- Relaying: For short ranged mmW communication, relaying could be a good option for BH technology.
- picocells and microcells: In city environments microcells are likely to be NLOS where as picocells might be LOS. It is likely that both would be significant [Z+13].
- Ownership: For better coverage mmW cells should be placed indoors [Z+13, NMSR13]. Similar to the femtocells [Ort08, CAG09, YTLK08, ACD+12], and neighborhood small cells [FKR13] concepts, it would be better if third party companies provide these cells.

Phase Noise

An ideal free running oscillator generates a signal

$$s(t) = Re(A_c e^{j2\pi f_c t + \phi_0}), \tag{4.106}$$

where A_c is the amplitude, f_c is the frequency, ϕ_0 is the initial phase, and $Re(z) = x$ denotes the real part of the complex number $z = x + jy$. For simplicity, we will assume $\phi_0 = 0$. Unfortunately, for practical oscillators, phase noise is introduced by an additional stochastic process $\phi(t)$ as $s(t) = Re(A_c e^{j2\pi f_c t + \phi(t)})$, whereas $\phi(t)$ is a realization of a Wiener process $\Theta(t)$ given as [GK13] $\Theta(t) = \Theta(0) + \int_{\tau=0}^{t} W(\tau)d\tau$, where $\Theta(0)$ is uniformly distributed between $-\pi$ and π and $W(t)$ satisfies $\mathbb{E}[W(t)] = 0$, where $\mathbb{E}[X]$ denotes expectation of X.

$$\mathbb{E}[W(t_1)W(t_2)] = 2\pi\beta\delta(t_2 - t_1) = \sqrt{c}\delta(t_2 - t_1), \tag{4.107}$$

where $\delta(t)$ is the Dirac-delta function, and the parameter β is called the full-width at half-maximum (FWHM), because the power spectral density of a Lorentzian shape, for which β is the FWHM, and $c = 4\pi^2\beta^2$. The variance of $\Theta(t)$ can be given as

$$Var[\Theta(t)] = ct. \tag{4.108}$$

Finally, the PSD of phase noise is defined as [SVTR09], $S(\Delta f) = \dfrac{c}{(2\pi\Delta f)^2 + (c/2)^2}$.

Most modern receivers use a type of phase-locked loop (PLL) for synchronizing with the Tx oscillator. In general, the PLL output is bound by the reference crystal oscillator (RCO) for frequencies below the loop bandwidth (LBW) and by the voltage controlled oscillator (VCO) for frequencies above LBW. Even for sub-6 GHz frequencies, VCOs are often plagued by flicker noise (FN) [PBL+13]. For a free running VCO, the variance of the combined phase and flicker noise is given by [SVTR09, Dem02] $\sigma^2(t) = (2\pi f_c)^2\{c_\omega t + c_f \int_0^t \int_0^t R_{Nm}(t_1 - t_2)dt_1 dt_2\}$, where c_ω and c_f are constants describing white noise and FN perturbations, respectively. The PSD of this white and FN is given as $L(\Delta f) = \dfrac{f_c^2(c_\omega + c_f S_f(\Delta f))}{\pi^2 f_c^4(c_\omega + c_f S_f(\Delta f))^2 + (\Delta f)^2}$, where f_c is the center frequency of oscillation and $S_f(\Delta f)$ is the PSD of FN.

Effect of Phase Noise in Single-Carrier Systems
Let us consider a system which transmits complex baseband symbols of duration T_s defined as [Kri15], $c(t) = \sum_{l=0}^{L-1} c_l p(t - lT_s)$, $p(.)$ is the transmitted pulse, and c_l is the complex symbol. The signal is upconverted and the bandpass representation of the transmitted signal is $c_{Tx}(t) = Re\{c(t)e^{j(2\pi f_c t + \phi_{Tx}(t))}\} + n(t)$, where $Re\{z\}$ is the real part of a complex number z, f_c is the carrier frequency, $\phi_{Tx}(t)$ is the phase noise added by the Tx local oscillator, and $n(t)$ is AWGN of mean 0 and variance $\dfrac{N_0}{2}$.

At Rx, the received signal is downconverted and filtered, the expression of the filtered signal $r(t) = c(t)e^{j\phi(t)} + n'(t)$, where $\phi(t) = \phi_{Tx}(t) + \phi_{Rx}(t)$, where $\phi_{Rx}(t)$ is the phase noise added by the Rx local oscillator, and $n'(t)$ is AWGN of mean 0 and variance N_0. If this signal is sampled at intervals kT_s, the expression for the sampled signal is $r(kT_s) = c_k e^{j\phi(kT_s)} + n'(kT_s)$, where $\phi(kT_s) = \phi((k-1)T_s) + \Delta_k$, where from (4.108), Δ_k is normal distributed with zero mean and variance $= cT_s$. The bit error rate of BPSK in the presence of phase noise is given by [CRY+11]

$$P_b = \frac{1}{2} \int_\Phi erfc(\sqrt{\frac{E_b}{N_0} cos(\phi)})p_\Phi(\phi)d\phi \qquad (4.109)$$

where $erfc$ is the complementary error function, $\frac{E_b}{N_0}$ is the ratio of bit energy to noise spectral density, and $p_\Phi(\phi)$ is the Gaussian probability density function of the residual phase jitter, which has a mean of 0 and variance of

σ_ϕ^2. In order to calculate σ_ϕ^2, first we consider the phase noise model [Spi77, HB94]

$$G_\phi(f) = \frac{k_a}{f^3} + \frac{k_b}{f^2} + k_c \qquad (4.110)$$

where the three terms in the right-hand side are representing the flicker noise, white frequency noise, and white phase noise respectively, and the constants k_a, k_b and k_c are determined by oscillator phase noise data.

The tracking error of the Rx PLL is given by [Spi77], $\sigma_p^2 = \int_0^\infty G_\phi(f)|1 - H(j\omega)|^2 df$, where $H(j\omega)$ is the closed loop transfer function of the second-order PLL. The tracking error of such a PLL with a damping factor of $\frac{1}{\sqrt{2}}$ is given as $|1 - H(j\omega)|^2 = \frac{(\frac{\omega}{\omega_n})^4}{1 + (\frac{\omega}{\omega_n})^4}$, where ω_n is the natural frequency. The above integration is evaluated and we obtain [Spi77] $\sigma_p^2 = \frac{8.71 K_a}{B_n^2} + \frac{3.7 K_b}{B_n^2} + K_c f_{max}$, where $B_n = 0.53\omega_n$ for a damping factor of $\frac{1}{\sqrt{2}}$. Also, the phase jitter performance of the carrier recovery circuit is given by [Feh81, HB94] $\sigma_t^2 = B_n T_b \left[\frac{1}{2R} + \frac{1}{4R^2}\right]$, where, T_b is the bit period and R is the SNR. The overall variance of the phase noise is $\sigma_\phi^2 = \sigma_p^2 + \sigma_t^2$

Effect of Phase Noise in Multi-Carrier Systems

An OFDM symbol in time domain is given by $x(n) = \sum_{k=0}^{N-1} s_k e^{j(2\pi/N)kn}$, where s_k is the complex data sequence and k denotes the kth subcarrier. This signal is frequency upconverted at the Tx and downconverted at the Rx, and the receiver output is given by (assuming a perfectly frequency flat channel) $s(n) = r(n) + w(n)$, where $r(n) = x(n)e^{j\phi(n)}$ and $w(n)$ is AWGN. $\phi(n)$ is the phase noise, modeled as a normally distributed random variable with mean 0 and variance σ_ϕ^2 as discussed in the previous section. Now, if $\phi(n)$ is small [Gar01], then we can write $e^{j\phi(n)} \approx 1 + j\phi(n)$. The demultiplexed signal is given as

$$y(k) \approx s_k + \frac{j}{N} \sum_{r=0}^{N-1} s_r \sum_{n=0}^{N-1} \phi(n) e^{\frac{2\pi n}{N}(r-k)}$$

$$\approx s_k + c_k \qquad (4.111)$$

If $k \neq r$, then we will have intercarrier interference which severely affects the performance of OFDM systems. For $k = r$, we will have a rotation of constellation by an angle Φ given as $\frac{1}{N}\sum_{n=0}^{N-1}\phi(n)$ [Gar01].

Non-linear Amplifier

An ideal PA is expected to satisfy the following relation between the input power (P_{in}) fed to it and the output power (P_{out}) obtained from it.

$$P_{out} = K_{gain} P_{in}, \qquad (4.112)$$

where K_{gain} is the power gain of the amplifier. However, real amplifiers do not follow such linear curves. For a practical PA, if P_{in} keeps increasing, the P_{out} starts to saturate after a point. If P_{in} is increased even further, the amplifier burns after a point. On the other hand, if the P_{in} is zero, the P_{out} is non-zero due to the noise generated in the PA [Poz05]. While specifying a microwave or an mmW PA, the following parameters are often mentioned

- Gain: The most obvious amplifier specification. Usually specified in dB.
- Peak output power: Peak output power (POP) is the highest point of a P_{in}–P_{out} curve of a PA. This usually decreases with the increase of frequency. So for mmW frequencies, since POP is lower as compared to microwave frequencies, the PA is often driven into saturation to get maximum possible output, sacrificing signal linearity to certain extent.
- AM/AM and AM/PM curves: Most of the modern digital modulation methods use complex constellations. Let us assume the input to a PA is the symbol, given in polar form as $re^{j\theta_r}$, and the corresponding output is $se^{j\theta_s}$. Then the AM/AM curve gives the relation

$$s = f_r(r) \qquad (4.113)$$

and AM/PM gives the relation

$$\theta_s = \theta_r + f_\theta(r, \theta_r) \qquad (4.114)$$

For an ideal amplifier, $f_r(r) = r\sqrt{K_{gain}}$ and $f_\theta(r, \theta_r) = 0$ for all values of r and θ_r.

- Dynamic Range: It is the range of P_{out} for which the PA has linear characteristics defined by (4.112)

Baseband Power Amplifier Models:

There are several different ways of modeling a microwave/mmW amplifier in baseband. Some of the popular models are

Saleh Model [Sal81]:   We assume that the input point of the constellation is $r_{in}e^{j\phi_{in}}$. The output constellation point $r_{out}e^{j\phi_{out}}$ is related by the

AM/AM response

$$r_{out} = \frac{\alpha_1 r_{in}}{1 + \beta_1 r_{in}^2}, \tag{4.115}$$

and AM/PM response

$$\phi_{out} = \phi_{in} + \frac{\alpha_2 r_{in}^2}{1 + \beta_2 r_{in}^2}. \tag{4.116}$$

Clearly for an ideal amplifier, $\beta_1 = \alpha_2 = 0$.

Ghorbani Model [GS91]: The AM/AM equation is given by

$$r_{out} = \frac{x_1 r_{in}^{x_2}}{1 + x_3 r_{in}^{x_2}} + x_4 r_{in}, \tag{4.117}$$

and AM/PM response

$$\phi_{out} = \phi_{in} + \frac{y_1 r_{in}^{y_2}}{1 + y_3 r_{in}^{y_2}} + y_4 r_{in}. \tag{4.118}$$

Rapp Model [Rap91]: The AM/AM equation is given by

$$r_{out} = \frac{r_{in}}{(1 + (\frac{r_{in}}{O_{sat}})^{2S})^{\frac{1}{2S}}}, \tag{4.119}$$

where S is the smoothness factor and O_{sat} is the output saturation level.

Passband Power Amplifier Model:
The most common passband PA model is given by

$$V_{out}(t) = K_1 V_{in}(t) + K_3 V^3{}_{in}(t) + K_5 V^5{}_{in}(t), \tag{4.120}$$

where k_1 is identified as the linear gain. Before deriving the high-order coefficient, we should convert IP3 (the 3 dB intercept point, i.e., where the output power of the third-order intermodulation product is equal to the output power of the desired signal) and P1dB (the 1 dB compression point) from power quantity to voltage quantity, as follows:

$$V_{IP3} = \sqrt{2 \times 10^{0.1(IP3-30)} \times R} \tag{4.121}$$

where R is system impedance. The third-order coefficient can be expressed as in [Tui02], $k_3 = \dfrac{4k_1}{3v_{IP3}^2}$. According to 1-dB compression point definition [36] and the fundamental distortion from third and fifth order [Tui02], the relationship between 1-dB compression and polynomial coefficients is:

$$\frac{k_1 v_{P1dB} + \frac{3}{4}k_3 v_{P1dB}^3 + \frac{5}{8}k_3 v_{P1dB}^5}{3v_{P1dB}^2} = 0.89125 \qquad (4.122)$$

The polynomial model might be inaccurate for higher input powers. The operation point of a HPA is usually determined by input backoff (IBO), which is the difference between the input power where the output power starts to saturate and the average input power usually given in dB.

Effect of ADC

The analog to digital conversion consists of three different processes, sampling, quantization, and coding [PM06]. The well-known quantization noise for sinusoidal input is given by

$$N_{quan} = \frac{\Delta^2}{12} \qquad (4.123)$$

where Δ is the quantization step size, given by [PM06]

$$\Delta = \frac{x_{max} - x_{min}}{L - 1} \qquad (4.124)$$

where x_{max} is the maximum value of the sampled signal, x_{min} is the minimum value of the sampled signal, and L is the number of quantization levels, given as $L = 2^N$, where N is the number of output bits of the quantizer. This additional quantization noise becomes important when AWGN variance is low [YXVG11].

A 1-bit ADC can be modeled as [MH14],

$$r = sgn(y), \qquad (4.125)$$

where y and r are input and output to the ADC, respectively, and sgn denotes the signum function given by,

$$sgn(x) = \begin{cases} 1, \text{for } x > 0 \\ 0, \text{for } x = 0 \\ -1, \text{for } x < 0. \end{cases} \qquad (4.126)$$

Similarly, the output of an N-bit ADC an analog signal between $x_{min} + (k - 0.5)\Delta$ to $x_{min} + (k + 0.5)\Delta$ is quantized to a level of $x_{min} + k\Delta$, where $0 \leq k \leq L - 1$ and x_{min}, Δ, and L are from (4.124) [PM06].

Effect of low-resolution ADCs:
The capacity of a single-input single-output (SISO) channel using QPSK with constellation of amplitude $\sqrt{P}$, 1-bit ADC, and a channel realization of $|h|\angle h$ completely known to both Tx and Rx is given by [MH14, SZM$^+$09],

$$C_{SISO}(P) = 2\left(1 - H\left(Q\left(|h|\sqrt{P}\right)\right)\right), \qquad (4.127)$$

where Q denotes Q-function and $H(p) = -[p\log_2 p + (1 - p)\log_2(1 - p)]$.

4.2.2 State-of-the-Art Technology and Standards

Hardware
Even a few years ago, the various components available at mmW frequencies were either very expensive or are inferior to their lower frequency counterparts. With the advent of very high f_T (current gain cutoff frequency), and f_{max} (maximum oscillation, frequency) now excellent amplifiers, oscillators, and mixers exist at mmWs [RMG11]. For example, [G$^+$10] reports the existence of heterojunction bipolar transistors (HBT) of f_T and f_{max} greater than 200 GHz; on the other hand, [H$^+$10] describes a terahertz microwave integrated circuit based on InP double HBT having an f_{max} of 808 GHz. The advancement of transistor technologies has been summarized in Figure 4.10.

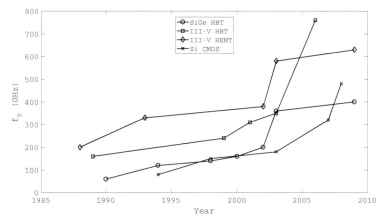

Figure 4.10 Evolution of microwave and millimeter wave transistors [RK09, H$^+$10, RMG11, G$^+$10].

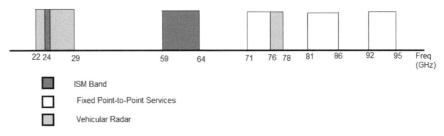

Figure 4.11   mmW frequency allocation in the USA [Y$^+$11].

An excellent summary of the development of millimeter wave components has been given in [RMG11]. Figure 4.11 shows the frequency allocation of mmW frequencies in the USA. There is 5 GHz available in the ISM bands between 59 and 64 GHz. The frequencies of 24 and 77 GHz are currently attached to automotive radar. Fixed point-to-point communication links can use frequencies of 71–76, 81–86 and 92–96 GHz which are licensed. Some of the uses of these bands are

- The large ISM bandwidth available around 60 GHz can be used for very high speed links of WPAN (IEEE 802.15.3c) or WLAN IEEE (802.11ad). The data rate can range from several hundred Mbps to a few Gbps.
- The 24 and 77 GHz frequencies are used in automotive and short ranged radar
- Licensed bandwidths at 71–76, 81–86 and 92–96 GHz can be used to design mmW links with datarates as high as 10 Gbps.

Figure 4.2 shows the air attenuation due to oxygen and water vapor for different frequencies. It can be seen that the oxygen attenuation is maximum at frequencies close to 60 GHz and hence these frequencies can be used only in short ranged communications. On the other hand, the frequencies in the vicinity of 28 GHz and 94 GHz have very low values of atmospheric attenuation, so they can be used for longer distance communications such as radio astronomy and mmW radars.

Another crucial aspect for mmW communications is rain attenuation. Usually, rain attenuation increases with increasing frequencies and often leads to the limit of the range of communications. System designers usually provide a significant margin for rain attenuation (depending on the long-term weather history of the area under consideration). However, for short ranges, rain attenuation is often less significant and for a microcell of a few hundred meters we can find a rain attenuation in the order of 2–3 dB for a heavy rainfall of 50 mm/h [GLR99].

Current Millimeter-Wave Standards
IEEE 802.15.3c:
An analysis by 802.15.3c task group (TG) for commercial applications in the 60 GHz band resulted in acceptance of five usage models (UMs) [B$^+$11, Sad06]. They are

- Uncompressed Video Streaming: The high bandwidth available at 60 GHz band should enable uncompressed high definition (HD) video streaming and eliminate the need of wires. The 802.15.3c TG calculated the required datarate to be around 3.5 Gb/s with a pixel error rate below 10^{-9} and range of 10 m.
- Uncompressed Multivideo Streaming: If the requirement is a single Tx transmitting multiple and different video streams to multiple Rx-s. According to this UM, this would require at least two 0.62 Gb/s streams.
- Office Desktop: According to this UM, the CPU of a computer should be able to do bidirectional communication with its peripherals wirelessly.
- Conference *ad hoc*: This UM would enable bidirectional ad hoc communications between computers in a network.
- Kiosk File Downloading: Using this UM, the subscriber should be able to download data and multimedia files from kiosks using low-power handheld devices. This requires a data rate of 1.5 Gbps and a range of 1 m.

Different countries allocated different frequencies around 60 GHz as unlicensed bands. So it is important to channelize these so that different countries can use different channels. The channelization is explained in Figure 4.12. Different countries can be allocated different channels as per their laws.

Physical Layer Design of IEEE 802.15.3c
In order to accommodate several conflicting UMs, the following standards have been proposed in 802.15.3c, and three different physical layer (PHY) models have been proposed [B$^+$11].

- Single-Carrier Mode (SC PHY)
- High-Speed Interface Mode (HSI PHY)
- Audio/Visual Mode (AV PHY)

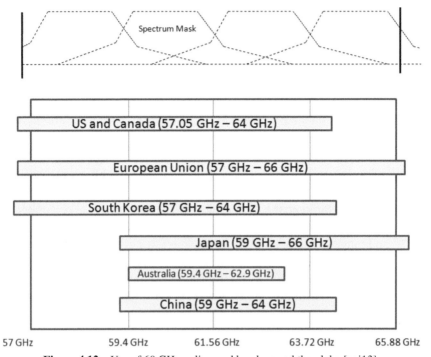

Figure 4.12   Use of 60 GHz unlicensed band around the globe [agi13].

SC PHY is best for UM3 and UM5, HSI PHY is suited where low latency is required such as UM4 and AV PHY is good for high datarate requirements such as UM1 and UM2.

MAC Layer Design of IEEE 802.15.3c

MAC Layer Design of IEEE 802.15.3c The MAC layer of IEEE 802.15.3c is derived from IEEE 802.15.3b [B+11]. In this standard, an ad-hoc piconet is formed between different devices. The role of piconet co-oridnator (PNC) is randomly assumed by one of the devices. PNC controls synchronization and access of all the DEVs. Upon receiving a signal from the PNC the DEV knows about the existence of the piconet.

During network operation, time is divided into sequential superframes (SFs). Each SF can be divided into a beacon period, a contention access period (CAP), and a channel time allocation period (CTAP). During the beacon period, the PNC sends beacons. Control and communication between the PNC and DEV is usually done in CAP period. Since the network is ad-hoc proper collision avoidance access scheme is used. The main data transmission using time division multiple access (TDMA) is done in CTAP period. So the already existing efficient MAC layer had only a few improvement requirements:

- Avoiding Interference.
- Improving transmission efficiency to enable MAC SAP rates over 1 Gb/s and providing low-latency transmissions for delay-sensitive applications.
- Supporting directivity inherent to 60-GHz signals and BF antennas.

IEEE 802.11ad

IEEE 802.11ad standard for Wireless LAN at 60 GHz was introduced in 2009 by a proposal from the Task Group "ad" (TGad) which was formed to suggest changes in the MAC and PHY layers of already existing IEEE 802.11 standard to support the so-called very high throughput (VHT) of at least 1 Gbps [agi13]. The proposal was accepted as TGad D0.1 in 2010. The standard supports short range communication (range upto 10 metres) with data rates upto 6.75 Gbps in 60 GHz. If required it can also switch to other ISM bands of 2.4 or 5 GHz.

802.11ad uses burst transmission mode that start with a single-carrier synchronization sequence followed by single or multicarrier header and payload data [agi13].

An excellent summary of physical layer aspects of IEEE 802.11ad protocol can be found in the application note [agi13].

IEEE 802.11ad Beamforming Protocol

A selection-based protocol is used for BF, which consists of three phases [KS16, 20112].

- Sector Level Sweep (SLS) Phase: In the SLS phase the Tx and Rx antennas use the so-called "quasi-omni" patterns, to select the most appropriate sector for Tx and optionally Rx antenna.
- Beam Refinement Phase (BNP): Once the best sector pairs have been selected, Tx and Rx antennas adjust themselves to more finer beamwidths.
- Beam Tracking (BT) Phase: This optional phase helps the Tx and Rx antennas to track any changes in the channel.

4.2.3 Millimeter-Wave Applications for 5G

One of the major objectives of 5G is 1000 times increase of datarates as compared to existing 4G systems. One of the methods to achieve this huge increase is by using ultra dense networks (UDN). UDN consists of much larger macrocells, which usually take care of the scheduling, allocation of resources and support for users in high speed vehicles, on the other hand small-cell BSs provide the faster data rate for low-mobility and mainly indoor users [GDM+15, IRH+14]. UDN is discussed in much greater detail in the next chapter. In order to provide such a huge amount of data, huge BH infrastructure is required, it has been shown, [TS15], that a BH bandwidth of 1 to 10 GHz is required to support an UDN. Fibre optic networks are reliable but they are not a very good choice for BH because restrictions in installation, placement and maintenance due to accessibility issues in remote areas. Therefore wireless BH is an interesting solution to this problem, notably in the mmW frequencies. Some of the advantages of wireless BH are

- Large amounts of underutilized networks: Most of the mmWave spectrum are still underutilized and lightly licensed.
- Smaller Antenna Dimensions: The antenna dimensions are proportional to the carrier signal wavelength, so in mmWave antenna dimensions are much smaller as compared to microwaves. So a much larger number of antennas can be packed for better directivity.

However, there are also certain challenges in mmWave, some of them are [GDM+15]

- More Expensive ADCs: The ADCs at mmWave frequencies are either of lower resolution or more expensive or both as compared to their microwave frequency counterparts. This forces us to do analog beam-forming.
- More difficult channel estimation: Since the number of antennas used at mmWave is much more than in microwaves channel estimation, which is crucial for multi user (MU-MIMO) and massive MIMO becomes more difficult even when using TDD.
- Necessity of channel state information at receiver: In microwave massive MIMO due to limited form factor, only channel state information at transmitter (CSIT) is required for precoding. However, in the case of mmWave Massive MIMO UDN each small cell is fixed with a large

number of antennas, so the CSIR is also required along with CSIT at macro/micro cell BS.

4.3 Rician K-Factor for Indoor mmWave Channels

It is mentioned above that 5G is expected to use mmW bands [A+14]. IEEE has already rolled out two mmW communication standards for 60-GHz indoor systems, IEEE 802.11ad [MML+11, YXVG11, agi13] for WLAN and IEEE 802.15.3c [B+11, YXVG11, 80209] for WPAN. In mmW bands, most of the power is carried by the LOS and the few earliest arriving clusters, and hence the communication is mainly LOS based.

Further, the performance of MIMO systems is significantly affected by channel characteristics such as correlation and LOS [IH09, ZVM10, HL16, JP03, KA06, PNG06]. LOS is characterized by the Rician K-factor [PNG06, Stu11]. We have also seen above that recent channel models like WINNER II [K+08], 3GPP [3gp17a], and COST 2100 [L+12] have provisions for the LOS component, which reminds us LOS conditions are important part of evaluation scenarios.

Evaluation of the above-mentioned systems require proper channel models such as 3GPP [3gp17a], its modified version for mmW bands [SR16], IEEE 802.11ad [agi13, YXVG11], and IEEE 802.15.3c [YXVG11]. They give an option to the user for adding a value of Rician K-factor to the first arriving ray. As an instance consider that the 3GPP model [3gp17a], suggests in the Urban Microcell LOS scenario to add Rician K-factor (in dB), is equal to $K = 13 - 0.03d$, where d is the distance between the Tx and the Rx to first path. It is also mentioned there that the Rician K-factor is correlated with several other channel parameters. However, such an extrinsic K-factor is often only a part of the actual Rician component experienced by a link. This is because the total NLOS power depends on several depends on cluster properties like the number of clusters and rays, their ToAs, AoAs, AoDs, and power azimuth spectrum (PAS). When the antenna gain and PAS of the clusters are considered for a particular link, then the effective value of K-factor perceived by the link may not be the same as the externally provided K-factor value. Further, the calculation of Rician K-factor is a function of cluster PAS, ToA, AoA, and AoD, which are all random variables. The effective Rician K-factor of a link is an important parameter as it affects several channel statistics such as probability density function (PDF), level crossing rate [Stu11], and spatial and temporal channel correlation of signal to noise ratio (SNR), which are necessary to arrive at the theoretical estimate of performance of a link such as bit error rate (BER), outage probability, channel capacity, etc. An analytical method to compute the effective value of Rician K-factor experienced by a link considering the Tx and Rx antenna gains, realistic channel propagation models which include cluster PAS, ToA, AoA, and AoD etc are summarized here.

4.3.1 Rician K-Factor Calculation for the Multicluster With Directivity Saleh–Valenzuela Model

In this work, we consider a typical system as shown in Figure 4.13. We assume that the Tx antenna is located at the origin, with its boresight pointed along the positive x-axis. The Rx antenna is located at $(R, 0, 0)$ with its boresight making an angle θ_0 with the positive z-axis and ϕ_0 with the negative x-axis. The signal from the Tx reaches the Rx through several clusters (only three of them are shown), rays from clusters arrive at Rx following PAS as prescribed by specific channel models. The Tx and Rx may use antennas with arbitrary antenna gain patterns. The channel model considered in this work is based on the original model by Saleh and Valenzuela [SV87], extended by Spencer *et al.* [SJJS00] for angular domain characteristics and Sawada *et al.*

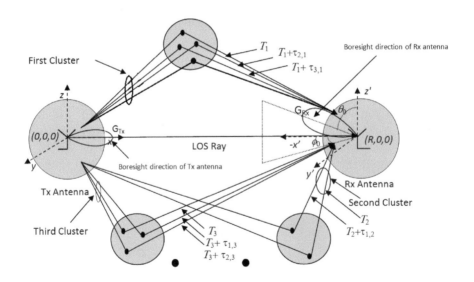

Figure 4.13   Antenna configuration [SM17].

[Saw06] for LOS component which is given as,

$$h(t, \tau, \phi_t, \theta_t, \phi_r, 0_r) = \sqrt{P_0}e^{j\kappa}\delta(t) + \beta(0,0)e^{j\alpha(0,0)}\delta(t)$$

$$+ \sum_{l=1}^{L}\sum_{k=0}^{K} \beta(k,l)e^{j\alpha(k,l)}$$

$$\times \delta(t - T_l - \tau_{k,l})\delta(\phi_r - \Phi_{r,l} - \phi_{r,k,l})$$

$$\times \delta(\phi_t - \Phi_{t,l} - \phi_{t,k,l})\delta(\theta_r - \Theta_{r,l} - \theta_{r,k,l})$$

$$\times \delta(\theta_t - \Theta_{t,l} - \theta_{t,k,l}), \tag{4.128}$$

where T_l is the ToA of the first ray of the lth cluster, $\tau_{k,l}$ is the difference of ToAs of the kth and first ray of the lth cluster, $\Theta_{r,l}$ and $\theta_{r,k,l}$ are the elevation AoA of the kth ray of the lth cluster $\Phi_{r,l}$ and $\phi_{r,k,l}$ are the azimuth AoA of the kth ray of the lth cluster, $\Theta_{t,l}$ and $\theta_{t,k,l}$ are the elevation AoD of the kth ray of the lth cluster, and $\Phi_{t,l}$ and $\phi_{t,k,l}$ are the azimuth AoD of the kth ray of the lth cluster. $\beta(k,l)$ and $\alpha(k,l)$ are the magnitude and phase of the channel gain of kth ray of the lth cluster, respectively. P_0 and κ are the externally applied deterministic quantities denoting the constant LOS term [Saw06].

The Rician K-factor is identified [PNG06, Stu11] as,

$$K = \frac{P_{LOS}}{P_{NLOS}}, \tag{4.129}$$

where P_{LOS} is the power in the LOS ray and P_{NLOS} is the sum of the powers of all the NLOS clusters.

To calculate the Rician K-factor, for (4.128) we first find the total average power gain P_{total}, which can be found by summing up the individual powers of all MPCs as,

$$P_{total} = \mathbb{E}_\chi\Big[\int_{\theta_t=0}^{\pi}\int_{\theta_r=0}^{\pi}\int_{\phi_t=-\pi}^{\pi}\int_{\phi_r=-\pi}^{\pi}\int_{\tau=0}^{\infty} |h(t,\tau,\phi_t,\theta_t,\phi_r,\theta_r)|^2$$

$$\times d\tau d\theta_r d\theta_t d\phi_r d\phi_t\Big]$$

$$= G_T(0,0)G_R(\theta_0,\phi_0)[P_0 + \overline{\beta^2(0,0)}] + \sum_{l=1}^{L}\sum_{k=0}^{K}\mathbb{E}_{\{T_l,\tau_{k,l}\}}[\overline{\beta^2(k,l)}]$$

$$\times \mathbb{E}_{\{\theta_{r,k,l},\phi_{r,k,l}|\Theta_{r,l},\Phi_{r,l}\}}[G_R(\theta_{r,k,l},\phi_{r,k,l})]$$

$$\times \mathbb{E}_{\{\theta_{t,k,l},\phi_{t,k,l}|\Theta_{t,l},\Phi_{t,l}\}}[G_T(\theta_{t,k,l},\phi_{t,k,l})], \tag{4.130}$$

where χ is given as $\chi = \{T_l, \tau_{k,l}, \{\theta_{r,k,l}, \phi_{r,k,l}|\Theta_{r,l}, \Phi_{r,l}\}, \{\theta_{t,k,l}, \phi_{t,k,l}|\Theta_{t,l}, \Phi_{t,l}\}\}$.

In (4.130), we consider ensemble mean over the random variables $\{\tau, \{\theta_{r,k,l}, \phi_{r,k,l}|\Theta_{r,l}, \Phi_{r,l}\}, \{\theta_{t,k,l}, \phi_{t,k,l}|\Theta_{t,l}, \Phi_{t,l}\}\}$. Since the CIR $(h(t, \tau, \phi_t, \theta_t, \phi_r, \theta_r))$ is ergodic in nature [Stu11], ensemble average is the same as time average hence we consider $\overline{\beta^2(0,0)}$ and $\overline{\beta^2(k,l)}$ as the time averaged power gains. The mean value of the power gain of the kth ray of the lth cluster follows the exponential rule (denoted by $\overline{\beta^2(k,l)}$) and is related to $\overline{\beta^2(0,0)}$ [SV87], as,

$$\overline{\beta^2(k,l)} = \overline{\beta^2(0,0)}e^{-\frac{\tau_{k,l}}{\gamma}}e^{-\frac{T_l}{\Gamma}}, \tag{4.131}$$

Using (4.131) in (4.130), we get,

$$P_{total} = P_0 G_T(0,0)G_R(\phi_0, \theta_0) + \overline{\beta^2(0,0)}G_T(0,0)G_R(\phi_0, \theta_0)$$

$$+ \overline{\beta^2(0,0)}\left(\sum_{l=1}^{L}\sum_{k=0}^{K}\mathbb{E}[G_R(\phi_{r,k,l} - \phi_0, \theta_{r,k,l} - \theta_0)]\right.$$

$$\times \mathbb{E}_{\{\theta_{r,k,l}, \phi_{r,k,l}|\Theta_{r,l}, \Phi_{r,l}\}}[G_T(\phi_{t,k,l}, \theta_{t,k,l})]\mathbb{E}_{\tau_{k,l}}[e^{-\frac{\tau_{k,l}}{\gamma}}]\mathbb{E}_{T_l}[e^{-\frac{T_l}{\Gamma}}]\right)$$

$$= P_{LOS} + P_{NLOS}, \tag{4.132}$$

where P_{LOS} is the LOS power, which can be computed as,

$$P_{LOS} = P_0 G_T(0,0)G_R(\phi_0, \theta_0). \tag{4.133}$$

P_{NLOS} is the combined total power all the L NLOS clusters, and can be computed as

$$P_{NLOS} = G_T(0,0)G_R(\phi_0, \theta_0)\overline{\beta^2(0,0)} + \overline{\beta^2(0,0)}$$

$$\times \left(\sum_{l=1}^{L}\sum_{k=0}^{K}\mathbb{E}_{\{\theta_{r,k,l}, \phi_{r,k,l}|\Theta_{r,l}, \Phi_{r,l}\}}[G_R(\phi_{r,k,l} - \phi_0, \theta_{r,k,l} - \theta_0)]\right.$$

$$\times \mathbb{E}_{\{\theta_{r,k,l}, \phi_{r,k,l}|\Theta_{r,l}, \Phi_{r,l}\}}[G_T(\phi_{t,k,l}, \theta_{t,k,l})]\mathbb{E}_{\tau_{k,l}}[e^{-\frac{\tau_{k,l}}{\gamma}}]\mathbb{E}_{T_l}[e^{-\frac{T_l}{\Gamma}}]\right). \tag{4.134}$$

Now, using (4.133) and (4.134) in (4.129), we can compute the K-factor. To compute the terms in (4.134), we require PAS of clusters as in [SJJS00] and [3gp17a]. The addition of the LOS component is described in [Saw06].

Further computations are detailed in [SM17]. The final expression for Rician K-factor becomes,

$$K_{Rician} = \frac{G_\phi(0)G_\phi(\phi_0)P_0}{\beta^2(0,0)[G_\phi(0)G_\phi(\phi_0) + \sum_{l=1}^{L}\sum_{k=0}^{K}(\frac{\Lambda\Gamma}{1+\Lambda\Gamma})^l(\frac{\lambda\gamma}{1+\lambda\gamma})^k}$$
$$\times \mathbb{E}_{\phi|\Phi_{t,l}}[G_\phi(\phi_{t,k,l})]\mathbb{E}_{\phi|\Phi_{r,l}}[G_\phi(\phi_{r,k,l} - \phi_0)]]$$
$$(4.135)$$

4.3.2 Rician K-Factor Calculation for the Modified SV Model for mmW

Some recent works have modified the original SV model to use it at mmW frequencies [LSE07, Saw06]. The section below describes the model and computes Rician K-factor for the same.

4.3.3 Modified SV Model for mmW

One such work is [LSE07], which modifies (4.131) as $20log(\beta_{k,l}) = n$, where n is a normal distributed random variable with mean μ and variance σ^2. In dB scale, $\beta_{k,l}$ is normal distributed, and hence $\beta_{k,l}$, in linear scale, is log-normal distributed. Therefore, $\beta_{k,l}^2 = 10^{0.1n}$; or $\beta_{k,l}^2 = e^{\xi n}$. The mean value is given as

$$\mathbb{E}_{\beta^2}[\beta_{k,l}^2] = \int_{n=0}^{\infty}\frac{1}{\sigma\sqrt{2\pi}}e^{-\frac{(ln(n)-\xi\mu)^2}{2\xi^2\sigma^2}}dn, \qquad (4.136)$$

where μ is a random variable given by [LSE07],

$$\mu = \frac{ln(\overline{\beta_{0,0}^2}) - \frac{T_l}{\Gamma} - \frac{\tau_{k,l}}{\gamma} - \frac{(\xi\sigma)^2}{2}}{\xi}. \qquad (4.137)$$

The values of $\mathbb{E}_{T_l}[e^{-\frac{T_l}{\Gamma}}]$ and $\sum_{k=0}^{K}\mathbb{E}_{\tau_{k,l}}[e^{-\frac{\tau_{k,l}}{\gamma}}]$ are calculated using the data for Γ, Λ, γ, and λ given in [LSE07]. Combining (4.136) and (4.137), we get,

$$\mathbb{E}_{\beta^2}[\beta_{k,l}^2] = \mathbb{E}_{\{T_l,\tau_{k,l}\}}[\int_{n=0}^{\infty}\frac{1}{\sigma\sqrt{2\pi}}e^{-\frac{(ln(n)-\xi\mu)^2}{2\xi^2\sigma^2}}dn] = \mathbb{E}_{\{T_l,\tau_{k,l}\}}[e^{\xi\mu+(\xi\sigma)^2/2}]$$
$$= e^{(\xi\sigma)^2/2}\mathbb{E}_{\{T_l,\tau_{k,l}\}}[e^{\xi\mu}] = \overline{\beta_{0,0}^2}\mathbb{E}_{T_l}[e^{-\frac{T_l}{\Gamma}}]\mathbb{E}_{\tau_{k,l}}[e^{-\frac{\tau_{k,l}}{\gamma}}]. \qquad (4.138)$$

We get the same result if we take the expectation of (4.131). Hence, approaching the same way, as in Section 4.3.1, we can calculate the new Rician K-factor to be the same as given in (4.135).

4.3.4 IEEE 802.11ad Model

Now we focus on the IEEE 802.11ad model, which is the standard for wireless LAN at 60 GHz [MML+11, agi13]. We use the same configuration of Tx-Rx antennas and the reflected clusters as in Section 4.3.1. For this model, there are three types of rays in a cluster. These are

- Pre-Cursor Rays: These are the rays which come before the arrival of the central ray in a cluster. The average amplitude of the kth ray of the lth cluster decays according to the following equation,

$$A_{f,k}(\tau_{k,l}) = A_f(0)e^{-\frac{|\tau_{k,l}|}{\gamma_f}}, \tag{4.139}$$

where $\tau_{k,l}$ is the difference of ToA of the central ray and the ray in question. $A_f(0)$ is given by,

$$K_f = 20log(|\frac{\alpha}{A_f(0)}|), \tag{4.140}$$

where α is the amplitude of the central ray and the values of α and K_f are available from measurements [MML+11].

- Central Ray: It is the ray with the highest average amplitude, denoted by α. The ToA of this ray is given to be T_l, which is the ToA of the cluster of which the ray is a part. The average amplitude of this ray is log-normal distributed.

- Post-Cursor Rays: These are the rays which come after the arrival of the central ray in a cluster. The average amplitude of the kth ray of the lth cluster decays according to the following equation,

$$A_{b,k}(\tau_{k,l}) = A_b(0)e^{-\frac{|\tau_{k,l}|}{\gamma_b}}, \tag{4.141}$$

where $\tau_{k,l}$ is the difference of ToA of the central ray and the ray in question. $A_b(0)$ is given by $K_b = 20log(|\frac{\alpha}{A_b(0)}|)$; just like pre-cursor rays, the value of K_b is available from measurements.

The power gain of the LOS cluster is given as,

$$P_{LOS} = G_T(0,0)G_R(\theta_0, \phi_0)(\frac{\lambda_c}{4\pi d})^2 P_0, \tag{4.142}$$

where λ_c is the wavelength corresponding to the central frequency and d is the length of the LOS path between Tx and Rx.

Now let us calculate P_{NLOS} enabling us to calculate the Rician K-factor as per (4.129). For each cluster, the total power is given by the sum of the powers of central ray and the pre- and post-cursor rays. We attempt to calculate them one by one.

The average power of the central ray of the lth NLOS cluster, for $d \gg R_l$, is given by,

$$P_{0,l} = \mathbb{E}_{g_l^2}[G_T(\Theta_{t,l}, \Phi_{t,l})G_R(\Theta_{r,l} - \theta_0, \Phi_{r,l} - \phi_0)(\frac{g_l \lambda_c}{4\pi d})^2], \quad (4.143)$$

where g_l is truncated log-normal distributed, such that $20log(g_l) \sim \mathcal{N}(\mu, \sigma^2) \Rightarrow ln(g_l^2) \sim \mathcal{N}(\xi\mu, (\xi\sigma)^2)$ and $g_l^2 \leq k.$ for all L clusters total

$$
\begin{aligned}
P_{NLOS} = &\sum_{l=1}^{L} \mathbb{E}_{g_l^2}[P_{0,l}] + \sum_{l=1}^{L} \sum_{k=1}^{N_f} [\mathbb{E}_{\{\theta_{t,k,l}, \phi_{t,k,l} | \Theta_{t,l}, \Phi_{t,l}\}} [G_T(\theta_{t,k,l}, \phi_{t,k,l})] \\
&\times \mathbb{E}_{\{\theta_{r,k,l}, \phi_{r,k,l} | \Theta_{r,l}, \Phi_{r,l}\}} \\
&\times [G_R(\theta_{r,k,l} - \theta_0, \phi_{r,k,l} - \phi_0)] \times [\mathbb{E}_{\{\tau_{k,l}, g_l^2\}} [A_{f,k}^2(\tau_{k,l})] \\
&+ \sum_{l=1}^{L} \sum_{k=1}^{N_b} [\mathbb{E}_{\{\theta_{t,k,l}, \phi_{t,k,l} | \Theta_{t,l}, \Phi_{t,l}\}} [G_T(\theta_{t,k,l}, \phi_{t,k,l})] \\
&\times \mathbb{E}_{\{\theta_{r,k,l}, \phi_{r,k,l} | \Theta_{r,l}, \Phi_{r,l}\}} [G_R(\theta_{r,k,l} - \theta_0, \phi_{r,k,l} - \phi_0)] \\
&\times [\mathbb{E}_{\{\tau_{k,l}, g_l^2\}} [A_{b,k}^2(\tau_{k,l})], \quad (4.144)
\end{aligned}
$$

where N_f and N_b are the number of pre-cursor and post-cursor rays, respectively, and $\Psi = \{\tau_{k,l}, \{\theta_{t,k,l}, \phi_{t,k,l} | \Theta_{t,l}, \Phi_{t,l}\}, \{\theta_{r,k,l}, \phi_{r,k,l} | \Theta_{r,l}, \Phi_{r,l}\}, g_l^2\}$. Further details can be found in [SM17]. We can use the values calculated in (4.142) and (4.144) in (4.129) to calculate the final value of the K-factor for IEEE 802.11ad model.

4.3.5 Variation of K-Factor With Respect to Link Orientations

We proceed to verify (4.129) calculated using (4.133) and (4.134) as a part to study the variation of K-factor with the link orientation. We also intend to bring out the effect of spatial variation of antenna boresight and cluster AoA/AoD orientation on K-factor. The setup is the same as given in Figure 4.13. The boresight direction of the Tx antenna is fixed along the positive x-axis, the number and AoA and AoD of clusters are kept the same as described earlier, while the boresight direction of the Rx antenna is changed,

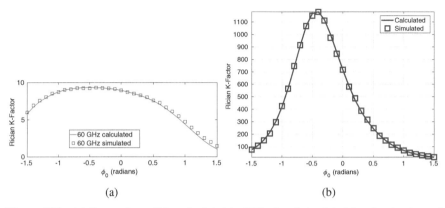

Figure 4.14 (a) Comparison of the calculated (solid line) and simulated (markers) values of Rician K-factor for SV-based channel model. (b) Comparison of the calculated (solid line) and simulated (squares) values of Rician K-factor for IEEE 802.11ad channel model [SM17].

which is measured by ϕ_0 on the xy plane. For the SV model, the variation of the Rician K-factor as a function of ϕ_0 is plotted in Figure 4.14(a). In the figure, the solid lines represent analytical results from (4.129), (4.133), and (4.134), whereas the markers show the simulated results. Good match between the simulated and calculated results is shown by the curves, and hence it can be concluded that the expressions derived in the previous section for SV-based models are usable for the theoretical performance evaluation of systems in such channels. From Figure 4.14(b), it can be inferred the K-factor is strongly influenced by the antenna configurations and cluster positions.

Now we show results with CIR generated following IEEE 802.11ad when the same experiments as the above are repeated. In Figure 4.18, the variation of both simulated and calculated (using (4.129), (4.142), and (4.144)), in markers) values of Rician K-factor with respect to ϕ_0 is plotted. An excellent match between the simulated and calculated results can be observed.

4.3.6 Impact of Rician K-Factor on System BER

It is shown above that spatial orientation of antenna and clusters affects the experienced K-factor in a link. Now it is demonstrated how this variation affects the performance of the communication system. Without loss of generality, we present results only for the SV model. Assumed value of Rician K-factor (considered to be $= \frac{P_0}{\beta^2(0,0)} = 10$ is often insufficient to determine the performance of a communication system using the channel.

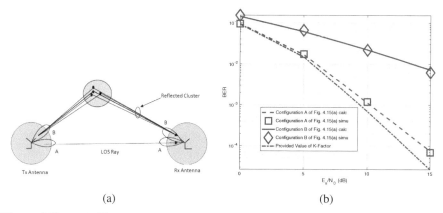

(a)   (b)

Figure 4.15   (a) Different configuration of Tx–Rx antennas giving rise to different values of K-factor. In configuration A, the boresights are aligned with the LOS ray, whereas in configuration B, the boresights are aligned with the cluster [SM17]. (b) BER for BPSK in different configurations for mmW [SM17].

Figure 4.15(a), configuration A, depicts the case, when the boresights of the Tx and Rx antennas are aligned with each other. We consider a single cluster with 10 rays having an AoA and AoD of 60°.

Since the NLOS rays are attenuated as they do not fall in the main lobe of the Tx and Rx antennas, the K-factor obtained in configuration A is much higher than that of configuration B. Figure 4.15(b), shows how BER is affected by inaccurate estimation of K-factor, for uncoded BPSK modulation.

For configuration A, an excellent match can be observed between the simulated bit error curve (shown in squares in Figure 4.15(b)) and the calculated BER curve (in dashed line). However, it can be noted that the BER plot for assumed K-factor (shown with dotted and dashed lines in Figure 4.15(b)) is somewhat away with the calculated and simulated K-factors. This is because of the difference between the provided and calculated values of the K-factor. Now if we consider, the results in configuration B, although the simulated results (shown in diamond markers) match well with the calculated results (shown in solid lines) they are significantly different from the results using assumed value of K-factor and the results for configuration A, because the changed configuration of antennas have significantly affected the K-factor value [SM17].

Next in Figure 4.15(a), configuration B, the Tx and Rx antennas are re-oriented, to make $\phi_0 = \phi_1 = 60^0$ so that the focus is toward the AoA and

AoD of the clusters along the boresight direction of the Tx and Rx antenna, respectively.

4.4 mmWave MIMO in Channels with Realistic Spatial Correlation

MIMO communications play a crucial role in today's wireless standards like LTE [HT09], WiFi (802.11n) [PO08], etc., and remain a major driving force for increasing SE of the next generation of mobile networks (5G). However, correlated spatial channel is known to degrade SE and BER performance [PNG03b] of MIMO systems. In this section, we describe the theoretical modeling of realistic MIMO spatial correlation of upcoming (5G) channel models and show its effectiveness in the performance evaluation of such future systems.

Practical MIMO channel models specified by 3GPP in [3gp15] and WIN-NER II [ATR16b, K+08] for sub-6 GHz and [HBK+16, TNMR14a, TR317] for mmWave have been defined for full dimensional MIMO systems and generate the channel coefficients based on various clustering parameters like PAS of arrival/departure, RMS angular spreads, angle bias, etc. These models are used in the performance evaluation of MIMO systems specified for LTE-Advanced and future 5G systems [3gp17a, 3gp17c, 3gp17b]. But such models only define the channel coefficients and do not provide analytically tractable expressions for spatial correlation. For realizing future 5G cellular systems, accurate characterization of the spatial correlation matrices for the above-mentioned channel models is necessary. It will help in reducing the time required for observing the channel performance metrics like SE, BER, etc., and will also help in deeper understanding of the channel behavior for different environments.

A MIMO system with polarized antennas both at the transmitter and at the receiver is considered. The antennas are arranged in a $(\sqrt{N} \times \sqrt{N})$ ($N = N_R$ or N_T, where N_R = number of receiver antennas, and N_T = number of transmitter antennas) UPA. The detailed realization of the 3GPP MIMO channel [3gp15] is given in [ATR16b]. This sections follows the above literature. The two-dimensional uniform planar antenna array is housed on the yz plane in the three-dimensional space. The transmitter and receiver are separated in 3D space by a distance d. If we fix the center of the transmitter UPA at the origin, then the center of the receiver UPA is located at (u_x, u_y, u_z), where

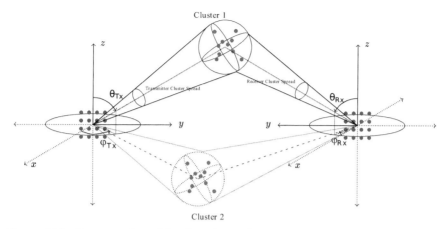

Figure 4.16 Representation of the clustering environment used in this work with UPA at both sides.

$\sqrt{u_x^2 + u_y^2 + u_z^2} = d$. A diagrammatic representation is given in Figure 4.16. Since the transmitter UPA has been taken at the origin, the coordinates of its antenna element z (2D index (p, q)) are $[0 \ (p - 1)d_y \ (q - 1)d_z]$. Similarly, for the receiver UPA, the coordinates of its antenna element u (2D index (c, d)) are $[u_x \ u_y + (c - 1)d_y \ u_z + (d - 1)d_z]$ (d_y and d_z are the horizontal and vertical inter-antenna spacings, respectively). The channel coefficient between the transmit antenna and receive antenna described above for the nth cluster (having a cluster power P_n and number of sub-rays M) is given as, [3gp15, Eq. 7.21]

$$H_{u,z,n}(t) = \sqrt{P_n} \sum_{m=1}^{M} \begin{bmatrix} F_{rx,u,V}(\theta_{rx,n,m}, \phi_{rx,n,m}) \\ F_{rx,u,H}(\theta_{rx,n,m}, \phi_{rx,n,m}) \end{bmatrix}^T$$

$$\begin{bmatrix} e^{j\psi_{n,m}^{vv}} & \sqrt{\kappa_{n,m}^{-1}}e^{j\psi_{n,m}^{vh}} \\ \sqrt{\kappa_{n,m}^{-1}}e^{j\psi_{n,m}^{hv}} & e^{j\psi_{n,m}^{hh}} \end{bmatrix} \begin{bmatrix} F_{tx,z,V}(\theta_{tx,n,m}, \phi_{tx,n,m}) \\ F_{tx,z,H}(\theta_{tx,n,m}, \phi_{tx,n,m}) \end{bmatrix}$$

$$\times e^{j2\pi[f_c t + \frac{u_x}{\lambda}\sin(\theta_{rx,m,n})\cos(\phi_{rx,m,n})]}$$

$$\times e^{j2\pi[\frac{u_y+(c-1)d_y}{\lambda}\sin(\theta_{rx,m,n})\sin(\phi_{rx,m,n})+\frac{u_z+(d-1)d_z}{\lambda}\cos(\theta_{rx,m,n})]}$$

$$\times e^{j2\pi[(p-1)\frac{d_y}{\lambda}\sin(\theta_{tx,m,n})\sin(\phi_{tx,m,n})+(q-1)\frac{d_z}{\lambda}\cos(\theta_{tx,m,n})]}.$$

$$(4.145)$$

The received signal at the uth antenna from the zth transmit antenna is given as,

$$r_{u,z}(t) = \sum_{n=1}^{N_C} H_{u,z,n}(t)s(t, \theta_{rx}, \phi_{rx}), \tag{4.146}$$

where N_C represents the number of clusters in the multi-cluster environment. $F_{rx,u,V}$ is the antenna gain for a particular RX antenna in the vertical polarization direction at the receiver at particular elevation and arrival angles. $\psi_{n,m}^{x,y}$ denote the phase difference between the y to x polarization components at the transmitter and receiver, respectively. The term $\kappa_{n,m}$ is the cross-polarization power ratio which indicates the ratio between the two polarization components: vertical and horizontal. The term θ_{rx} represents a variable denoting the elevation angles of the clusters at the receiver while the term ϕ_{rx} represents the variable denoting the azimuth angles at the receiver. The terms θ_{tx} and ϕ_{tx} represent the same variables at the transmitter. Terms $\theta_{rx,m,n}$ and $\phi_{rx,m,n}$ denote the elevation and azimuth subpath angles (m^{th} subpath of the nth cluster) at the receiver, respectively. $s(t, \theta_{rx}, \phi_{rx})$ represents the signal impinging on the receiver antenna array from the far field at a particular time instant t which follows the assumption made in [SR05]. If we take the average energy of the signal over time at a particular angular direction at the receiver, we get the PAS at the receiver. It is to be noted that this impinging signal is not directly related to the transmitted signal. The signal power which is transmitted is only received partially. The power which is not received is captured in the large-scale propagation losses which do not play any role in the calculation of spatial correlation as they do not depend on the spatial directions. The antenna gains are obtained from the 3GPP standard and depend on the polarization angle. They are given as [3gp15, Section 7.1.1],

$$F_{tx,V}(\theta_{tx}, \phi_{tx}) = \sqrt{A'(\theta_{tx}, \phi_{tx})}cos(\zeta) \tag{4.147}$$
$$F_{tx,H}(\theta_{tx}, \phi_{tx}) = \sqrt{A'(\theta_{tx}, \phi_{tx})}sin(\zeta), \tag{4.148}$$

and,

$$F_{rx,V}(\theta_{rx}, \phi_{rx}) = \sqrt{A'(\theta_{rx}, \phi_{rx})}cos(\zeta) \tag{4.149}$$
$$F_{rx,H}(\theta_{rx}, \phi_{rx}) = \sqrt{A'(\theta_{rx}, \phi_{rx})}sin(\zeta), \tag{4.150}$$

where from [3gp15, Table 7.1-1],

$$A'(\theta_{rx/tx}, \phi_{rx/tx})[dB] = -min\{-[A_{E,V}(\phi_{rx/tx}) + A_{E,H}(\theta_{rx/tx})], A_m\}, \tag{4.151}$$

with,

$$A_{E,V}(\phi_{rx/tx}) = -min\left[12\left(\frac{\phi_{rx/tx} - \frac{\pi}{2}}{\phi_{3dB}}\right)^2, SLA_v\right] \qquad (4.152)$$

$$A_{E,H}(\theta_{rx/tx}) = -min\left[12\left(\frac{\theta_{rx/tx}}{\theta_{3dB}}\right)^2, A_m\right], \qquad (4.153)$$

where SLA_v and A_m are the gains for the flat part of the antenna gain pattern in dB domain for the vertical and horizontal polarization components, respectively. The term ζ is the polarization angle where $\pi/4$ is cross-polarized, and $0, \pi/2$ are horizontally and vertically polarized cases, respectively. The terms $\theta_{rx/tx}$ and $\phi_{rx/tx}$ are also variables used to denote the same θ_{rx} and θ_{tx} (or ϕ_{rx} and ϕ_{tx}) as defined earlier but written in a consolidated format. It means that a common notation has been used to define certain quantities at the transmitter or the receiver.

P_n is the cluster power for the nth cluster and is generated from the temporal domain as [3gp15, Eq. 7.7],

$$P_n = \frac{\exp\left(-\tau_n \frac{r_\tau - 1}{r_\tau \sigma_\tau}\right) 10^{\frac{-Z_n}{10}}}{\sum_{n=1}^{N_C} \exp\left(-\tau_n \frac{r_\tau - 1}{r_\tau \sigma_\tau}\right) 10^{\frac{-Z_n}{10}}}, \qquad (4.154)$$

where r_τ is the delay scaling factor, σ_τ is the RMS delay spread, and Z_n is the per-cluster shadowing factor which is a log-normal random variable with mean 0.

4.4.1 Analytical Modeling of the Composite PAS

The cluster mean angles are obtained from the cluster powers using inverse Gaussian and Laplacian functions for the azimuth and elevation domains, respectively, following 3GPP specifications [3gp15]. The relationship between the mean cluster angles and the cluster powers has been given as [3gp15, Eqs. 7.8, 7.13],

$$\phi'_{n,AoA} = \frac{2(\sigma_{ASA}/1.4)\sqrt{-ln(P_n/(P_{max}))}}{C_{az}}. \qquad (4.155)$$

For azimuth and for elevation, the mean cluster angle is obtained as,

$$\theta'_{n,ZoA} = \frac{\sigma_{ZSA}ln(P_n/(P_{max}))}{C_{ele}}, \qquad (4.156)$$

where P_{max} denotes the maximum cluster power, and C_{az} and C_{ele} are scaling factors which depend on the number of clusters as described in [3gp15, Section 7.3]. An additional shift by the LOS angle is suggested so that the composite PAS is centered around LOS. Hence, the mean cluster angles become [3gp15, Eqs. 7.10,7.15],

$$\phi_{n,AoA} = sgn(\mathcal{U}(-1,1))\phi'_{n,AoA} + \phi_{rx,LOS} + \mathcal{N}\left(0, \frac{\sigma^2_{ASA}}{49}\right), \quad (4.157)$$

and,

$$\theta_{n,ZoA} = sgn(\mathcal{U}(-1,1))\theta'_{n,ZoA} + \theta_{rx,LOS} + \mathcal{N}\left(0, \frac{\sigma^2_{ZSA}}{49}\right) + \theta_{bias}, \quad (4.158)$$

where $\mathcal{N}(\mu, \sigma^2)$ is a normally distributed random variable with mean μ and variance σ^2, $\mathcal{U}(a, b)$ is a uniformly distributed random variable between a and b, and $sgn()$ denotes the sign of the quantity inside the brackets. So, in the azimuth, the cluster powers are distributed as Gaussian centered around $\phi_{Rx/Tx,LOS}$ and in elevation, they are distributed as Laplacian centered around $\theta_{Rx/Tx,LOS}$. Therefore, the individual cluster powers in the azimuth domain are given as,

$$P_{n,az} = P_{max}e^{-\left(\frac{\phi_{rx/tx,n} - \phi_{Rx/Tx,LOS}}{2\sigma_{ASA/ASD}}\right)^2 \frac{1}{1.4C_{az}}}, \quad (4.159)$$

and the elevation cluster power is given as,

$$P_{n,ele} = P_{max}e^{-\frac{|\theta_{rx/tx,n} - \theta_{Rx/Tx,LOS} - \theta_{bias}|}{\sigma_{ZSA/ZSD}} \frac{1}{C_{ele}}}. \quad (4.160)$$

The cluster power is given as $P_n = \sqrt{P_{n,az} \times P_{n,ele}}$. The quantities $\sigma_{ASA/ASD}$ and $\sigma_{ZSA/ZSD}$ denote the total RMS angular spreads in the azimuth and elevation domains. The terms $\theta_{Rx/Tx,LOS}$ and $\phi_{Rx/Tx,LOS}$ are the LOS angles in the elevation and azimuth directions, respectively. The LOS angles are calculated based on the position of the line joining the centers of the Tx and Rx antenna arrays with respect to the coordinate systems at the both ends and θ_{bias} is the LOS elevation bias angle as defined in [3gp15, Section 7.3]. In Figure 4.17, the LOS direction between the transmitter and the receiver has been represented and the LOS angles have been shown in the coordinate

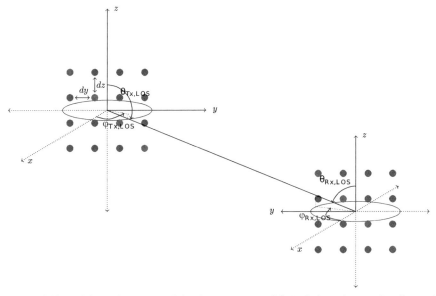

Figure 4.17 LOS angles assumed in the system model (red dots denote the diagonal elements of the UPA).

system. The individual cluster PAS is a product of the azimuth and elevation PAS of the cluster because the elevation and azimuth domains are stated to be independent in [3gp15]. The per-cluster PAS is distributed uniformly about the mean cluster angles. The width of the cluster PAS is proportional to the cluster angle spread. Therefore, the per-cluster PAS in 3D can be written as,

$$\text{PAS}_n(\Omega_{rx}) = \frac{\sqrt{P_{n,az}}}{2\alpha c_{\text{ASA}}} \times \frac{\sqrt{P_{n,ele}}}{2\alpha c_{\text{ZSA}}},$$
$$\phi_{rx,n} - \alpha c_{\text{ASA}} \le \phi_{rx} \le \phi_{rx,n} + \alpha c_{\text{ASA}},$$
$$\theta_{rx,n} - \alpha c_{\text{ZSA}} \le \theta_{rx} \le \theta_{rx,n} + \alpha c_{\text{ZSA}}, \qquad (4.161)$$

where $2\alpha c_{\text{ASA}}$ and $2\alpha c_{\text{ZSA}}$ are the per-cluster rms angular spread in the azimuth and elevation domains at the receiver. The variable $\alpha = 2.15$ for sub-6-GHz bands [Section 7.3] [3gp15] and $\alpha = 1.5$ for 28-GHz bands [TNMR14a]. The receiver side correlation between two antennas u[Index (p, q)] and u'[Index $(p1, q1)$] is calculated assuming a single transmit antenna z. The correlation between two antennas at the receiver is calculated by taking the expectation of $r_{u,z}(t)r^*_{u',z}(t)$ over all parameters like AoAs in elevation and azimuth domains, polarization phase difference, and cross-polarization power ratio, all of which are random variables [3gp15, DNI$^+$11]. The primary definition for the spatial correlation is shown in (4.162) and the final expression is shown in (4.164). The term G_u in (4.163) can be written as,

$$
\rho_{u,u'} = \frac{\mathbb{E}_{t,\kappa,\psi,\Omega_{rx}}[r_{(u,z)}(t,\theta_{rx},\phi_{rx})r^*_{(u',z)}(t,\theta_{rx},\phi_{rx})]}{\underbrace{\sqrt{\mathbb{E}_{t,\kappa,\psi,\Omega_{rx}}(|r_{u,z}(t,\theta_{rx},\phi_{rx})|^2)}\sqrt{\mathbb{E}_{t,\kappa,\psi,\Omega_{rx}}(|r^*_{u',z}(t,\theta_{rx},\phi_{rx})|^2)}}_{D}}
$$

$$
- \frac{\mathbb{E}_{t,\kappa,\psi,\Omega_{rx}}[r_{(u,z)}(t,\theta_{rx},\phi_{rx}))]\mathbb{E}_{t,\kappa,\psi,\Omega_{rx}}[(r^*_{(u',z)}(t,\theta_{rx},\phi_{rx}))]}{}
$$

(4.162)

$$
= \frac{\mathbb{E}_{t,\kappa,\psi,\Omega_{rx}}\left[\sum_{n=1}^{N_c}\sum_{m=1}^{M}G_u G^*_{u'}|s(t,\theta_{rx,m,n},\phi_{rx,m,n})|^2 \times e^{j2\pi[(p-p1)\frac{d_y}{\lambda}sin(\theta_{rx,m,n})sin(\phi_{rx,m,n})+(q-q1)\frac{d_z}{\lambda}cos(\theta_{rx,m,n})]}\right]}{D}
$$

(4.163)

$$
= \frac{\sum_{n=1}^{N_c}\int_{\theta_{rx}}\int_{\phi_{rx}}F_{u,u'}(\theta_{rx},\phi_{rx}) \times e^{j2\pi[(p-p1)\frac{d_y}{\lambda}sin(\theta_{rx})sin(\phi_{rx})+(q-q1)\frac{d_z}{\lambda}cos(\theta_{rx})]}\text{PAS}_n(\theta_{rx}) \times \text{PAS}_n(\phi_{rx})sin(\theta_{rx})d\phi_{rx}d\theta_{rx}}{D}
$$

(4.164)

$$
G_u = \begin{bmatrix} F_{rx,u,V}(\theta_{rx,n,m},\phi_{rx,n,m}) \\ F_{rx,u,H}(\theta_{rx,n,m},\phi_{rx,n,m}) \end{bmatrix}^T \times \begin{bmatrix} e^{j\psi^{vv}_{n,m}} & \sqrt{\kappa^{-1}_{n,m}}e^{j\psi^{vh}_{n,m}} \\ \sqrt{\kappa^{-1}_{n,m}}e^{j\psi^{hv}_{n,m}} & e^{j\psi^{hh}_{n,m}} \end{bmatrix}
$$
$$
\begin{bmatrix} F_{tx,s,V}(\theta_{tx,n,m},\phi_{tx,n,m}) \\ F_{tx,s,H}(\theta_{tx,n,m},\phi_{tx,n,m}) \end{bmatrix}.
$$

(4.165)

The expectation of $G_u G_{u'}$ over the random parameters inside the terms, i.e., κ and ψ has been obtained as $F_{u,u'}(\theta_{rx,n,m},\phi_{rx,n,m})$ in (4.166). The summation over the subrays and the clusters is replaced by integration over the solid angle when taking expectation over the angular domain. The term $\mathbb{E}_\kappa[\kappa^{-1}_{n,m}]$

is calculated by knowing that κ is log-normally distributed

$$\mathbb{E}_{\kappa,\psi|\Omega_{rx}}[G_u G_{u'}^*]$$

$$= \mathbb{E}_{\kappa,\psi|\Omega_{rx}} \left[\begin{bmatrix} F_{rx,u,V}(\theta_{rx,n,m}, \phi_{rx,n,m}) \\ F_{rx,u,H}(\theta_{rx,n,m}, \phi_{rx,n,m}) \end{bmatrix}^T \right.$$

$$\times \begin{bmatrix} e^{j\psi_{n,m}^{vv}} & \sqrt{\kappa_{n,m}^{-1}} e^{j\psi_{n,m}^{vh}} \\ \sqrt{\kappa_{n,m}^{-1}} e^{j\psi_{n,m}^{hv}} & e^{j\psi_{n,m}^{hh}} \end{bmatrix}$$

$$\times \begin{bmatrix} F_{tx,s,V}(\theta_{tx,n,m}, \phi_{tx,n,m}) \\ F_{tx,s,H}(\theta_{tx,n,m}, \phi_{tx,n,m}) \end{bmatrix} \begin{bmatrix} F_{rx,u',V}(\theta_{rx,n,m}, \phi_{rx,n,m}) \\ F_{rx,u',H}(\theta_{rx,n,m}, \phi_{rx,n,m}) \end{bmatrix}^T$$

$$\left. \times \begin{bmatrix} e^{-j\psi_{n,m}^{vv}} & \sqrt{\kappa_{n,m}^{-1}} e^{-j\psi_{n,m}^{vh}} \\ \sqrt{\kappa_{n,m}^{-1}} e^{-j\psi_{n,m}^{hv}} & e^{-j\psi_{n,m}^{hh}} \end{bmatrix} \begin{bmatrix} F_{tx,s,V}(\theta_{tx,n,m}, \phi_{tx,n,m}) \\ F_{tx,s,H}(\theta_{tx,n,m}, \phi_{tx,n,m}) \end{bmatrix} \right]$$

$$= \mathbb{F}_{u,u'}(\theta_{rx,n,m}, \phi_{rx,n,m}) \tag{4.166}$$

with mean and variance $(\mu_{XPD}, \sigma_{XPD})$. Moreover, from the definition of PAS, it is true that $\mathbb{E}_{t|\Omega_{rx}}(|s(t, \theta_{rx}, \phi_{rx})|^2) = \sum_{n=1}^{N_c} \text{PAS}_n(\theta_{rx})\text{PAS}_n(\phi_{rx})$.

4.4.2 Performance Analysis

All the following MIMO point-to-point link results have been produced using the cluster parameters from [HBK+16, Table V]. Urban Micro scenario has been considered with multiple clusters and in NLOS propagation conditions, where inter-element spacing in the antenna array (UPA) $\lambda/4$ both in horizontal and vertical directions. A brief overview of the system parameters is summarized in Table 4.1.

In Figure 4.18(a), the correlation between two antenna elements along the diagonal of a UPA has been plotted with respect to their separation in terms of multiples of the wavelength. The inter-antenna spacings along the diagonal are taken as multiples of $\frac{\lambda}{4}$. We see that the analytical correlation values generated from (4.164) almost match with the simulated results for most antenna spacings. We also see that the transmitter side correlation is higher than the receiver. This is because the transmitter side has a lower angular spread as evident from the data in Table 4.1.

Table 4.1   Large-scale parameters

	28 GHz
No. of clusters	6
No. of subpaths per cluster	10
BS antenna height	22
UE antenna height	5
RMS delay spread (ns)	129.68
RMS azimuth angle of arrival spread (^{0})	21.4716
RMS azimuth angle of departure spread (^{0})	3.1875
RMS elevation angle of arrival spread (^{0})	3.7130
RMS elevation angle of departure spread (^{0})	2.8437
Cluster azimuth angle of arrival spread (^{0})	6.8
Cluster azimuth angle of departure spread (^{0})	4.8
Cluster elevation angle of arrival spread (^{0})	6.4
Cluster elevation angle of departure spread (^{0})	2.2
Elevation angle of departure bias (^{0})	−7.6913
Elevation angle of arrival bias (^{0})	8.0334
LOS azimuth angle (^{0})	90
LOS elevation angle (^{0})	90
Tx-Rx distance (metres)	100
Cross-polarization power ratio (XPD)	26.3162

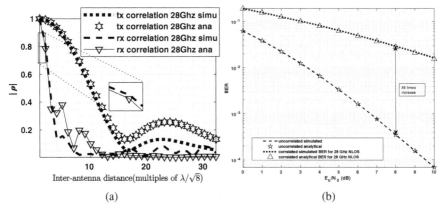

(a)   (b)

Figure 4.18   (a) Correlation coefficient across a UPA at the receiver and the transmitter. (b) BER performance of MRC system with four receiver antennas.

Next, in Figure 4.18(b), we observe the BER performance of an MRC system with BPSK for $N_T = 1$ and $N_R = 4$. Here, naturally only receiver side correlation matrix is involved. For calculating the BER, we use the expression as given in [PNG03b, (5.5)]. The BER for one instantaneous channel matrix $\mathbf{H}_{M \times 1}$ is upper bounded by,

$$P_e \leq \bar{N}_e \exp\left(-\frac{\eta d_{min}^2}{4} \right),$$ (4.167)

where $\eta = \sum_{i=1}^{N} |h_i|^2 \rho$, $\rho = \frac{E_b}{N_0}$, where $\frac{E_b}{N_0}$ is the symbol to noise power ratio. The theoretical average probability of the error upper bound $\bar{P}_e$ is given by [(5.6)] [PNG03b], $\bar{P}_e \leq \bar{N}_e \prod_{i=1}^{M} \frac{1}{1+\rho d_{min}^2 \lambda_i(\mathbf{R})/4}$, where d_{min} is the minimum distance between two points in the constellation used ($d_{min} = 2$ for BPSK) and $\bar{N}_e$ denotes the number of nearest neighbors in the constellation ($\bar{N}_e = 1$ here). The symbol $\lambda_i(\mathbf{R})$ is the ith eigenvalue of the channel covariance matrix $\mathbf{R}$. As per the Kroenecker model given in [PNG03b, (3.26)], it can be expressed as, $\mathbf{R} = \mathbf{R_t}^T \otimes \mathbf{R_r}$, where $\mathbf{R_r}$ is the receiver and transmitter correlation matrices, respectively, and obtained from (4.164). Therefore, it can be seen that the average error probability upper bound can be calculated from the correlation matrix directly. From the figure, it can be observed that, as expected, correlation degrades the BER performance significantly (almost by $> 10^2$ at 10 dB). Also, it is seen that all the analytical BER values match closely with the simulated ones. Therefore, the upper bound of the average P_e can be obtained directly from the correlation matrices without any time-consuming simulation for the mmWave channel using the methods described above.

The link SE for a spatial multiplexing system with the channel unknown to the transmitter is obtained from [PNG03b, Section 4.3].

$$C_{Ukn} = \sum_{i=1}^{r} \log_2(1 + \frac{\rho \lambda_i}{N_T}),$$ (4.168)

and that with the channel known to the transmitter is,

$$C_{Kn} = \sum_{i=1}^{r} \log_2(1 + \frac{\rho \lambda_i \gamma_i}{N_T}),$$ (4.169)

where the rank of the channel matrix is $r \leq \min(N_R, N_T)$ and λ_i's are the eigen values of $\mathbf{HH}^{\mathbf{H}}$ where $\mathbf{H}$ is the channel matrix. For the simulated channel, we calculate the λ_i's directly from the channel matrix $\mathbf{H}$, i.e., $\lambda_i = eig_i(\mathbf{HH}^{\mathbf{H}})$. To generate the analytical channel matrix, we use the Kroenecker model as follows,

$$\mathbf{H} = \mathbf{R_r}^{1/2} \mathbf{H_w} \mathbf{R_t}^{1/2},$$ (4.170)

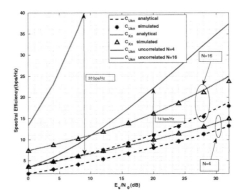

Figure 4.19   Comparison of SEs for polarized UPA of different dimensions in 28-GHz channel environments.

where the matrices $\mathbf{R_r}$ and $\mathbf{R_t}$ are generated from (4.164). The SE is calculated in the same way as for the simulated case. When the channel is known at the transmitter, additional water-filling power coefficients γ_i are used as defined in [PNG03b, (4.24)].

Figure 4.19 shows the SE with $N_R = N_T = 4, 16$. It is seen that the simulated SE matches with the analytical results satisfactorily for all the cases. It is also inferred that since the mmWave channel at the 28-GHz band has less number of clusters with small cluster angular spread, the correlation is in general large spectral efficiency loss. In fact, we can observe an SE gap of around 14 bps/Hz for four antennas at 20 dB SNR and around 33 bps/Hz for 16 antennas at only 9 dB SNR for the channel unknown at the transmitter case. From this, one is brought to the reality of achievable capacity in mmWave MIMO configurations. The huge capacity loss due to spatial correlation is found to affect with more ferocity as the number of antennas increases. The design of MIMO communication systems in mmWave thus requires more careful attention than otherwise. Finally, it is also important to note that using the method described above one can easily estimate the performance of MIMO scheme in a realistic spatially correlated MIMO channel without employing huge time- and power-consuming simulation facility yet achieve near perfect performance estimation.

4.4.3 Hybrid Beamforming

In the previous section, it is found that spatial correlation severely affects MIMO channel capacity, especially as the number of antennas increases. In

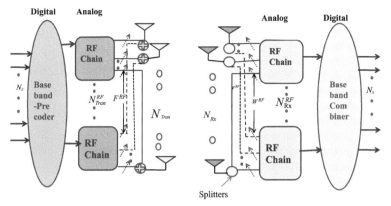

Figure 4.20 Transceiver design for hybrid BF.

mmWave, communication systems, there is a huge opportunity to a large number of antennas owing to the small wavelength. However, as the capacity loss is significant, while the cost of implementing individual RF section for each antenna increases cost and power consumption, a different approach is used. BF is a preferred method for using multiple antennas in such frequencies. This was also described in the earlier sections. Among the different BF techniques, a prominent architecture known as hybrid beam forming (HBF) [TG07] is becoming popular. The HBF is architecture is described below using Figure 4.20. Suppose we take into consideration a single-user mmWave MIMO system which has N_T antennas at the transmitter and communicates N_s data streams to a receiver with N_R antennas. Moreover, the transmitter is equipped with $N_{T,RF}$ transmit RF chains such that $N_s \leq N_{T,RF} \leq N_T$ [EARAS$^+$14]. Hence, the transmitter applies an $N_{T,RF} \times N_s$ baseband precoder $\mathbf{F_{BB}}$ using its $N_{T,RF}$ transmit RF chains, followed by an $N_T \times N_{T,RF}$ RF precoder $\mathbf{F_{RF}}$ using analog circuitry as shown in the figure below.

The discrete-time transmitted signal is given by,

$$\mathbf{x} = \mathbf{F_{RF}}\mathbf{F_{BB}}\mathbf{s} \tag{4.171}$$

where $\mathbf{s}$ is the $N_s \times 1$ symbol vector. Since $\mathbf{F_{RF}}$ is implemented using analog phase shifters, its elements are constrained to satisfy $(\mathbf{F_{RF}}^i\mathbf{F_{RF}}^{i^H})_{l,l} = N_T^{-1}$). The transmitter's total power constraint is enforced by normalizing $\mathbf{F_{BB}}$ such that $\|\mathbf{F_{RF}}\mathbf{F_{BB}}\|_F^2 = N_s$. Considering a narrowband block-fading propagation channel, the received signal can be written as [EARAS$^+$14],

$$\mathbf{y} = \sqrt{\rho}\mathbf{H}\mathbf{F_{RF}}\mathbf{F_{BB}}\mathbf{s} + \mathbf{n} \tag{4.172}$$

where $\mathbf{y}$ is the $N_R \times 1$ received vector, $\mathbf{H}$ is the $N_R \times N_T$ channel matrix such that $\mathbb{E}||\mathbf{H}||_F^2 = N_T N_R$, ρ represents the average received power, and $\mathbf{n}$ is the vector of i.i.d $\mathbb{C}N(0, \sigma_n^2)$ noise. The receiver uses its $N_s \leq N_{R,RF} \leq N_R$ RF chains and analog phase shifters to obtain the processed received signal.

$$\tilde{\mathbf{y}} = \sqrt{\rho}\mathbf{W_{BB}}^H\mathbf{W_{RF}}^H\mathbf{H}\mathbf{F_{RF}}\mathbf{F_{BB}}\mathbf{s} + \mathbf{W_{BB}}^H\mathbf{W_{RF}}^H\mathbf{n} \qquad (4.173)$$

where $\mathbf{W_{RF}}$ is the $N_R \times N_{R,RF}$ RF combining matrix and $\mathbf{W_{BB}}$ is the $N_{R,RF} \times N_s$ baseband combining matrix. Similar to the RF precoder, $\mathbf{W_{RF}}$ is implemented using phase shifters and therefore is such that $(\mathbf{W_{RF}}^i\mathbf{W_{RF}}^{iH})_{l,l} = N_R^{-1})$.

In order to maximize the SE, at the transmitter, we design $\mathbf{F_{RF}}\mathbf{F_{BB}}$ to maximize the mutual information achieved by Gaussian signaling over the mmWave channel [EARAS+14],

$$\mathbb{I}(\mathbf{F_{RF}}, \mathbf{F_{BB}}) = log_2\left(\left|\mathbf{I} + \frac{\rho}{N_s \sigma_n^2}\mathbf{H}\mathbf{F_{RF}}\mathbf{F_{BB}}\mathbf{F_{BB}}^H\mathbf{F_{RF}}^H\mathbf{H}^H\right|\right) \qquad (4.174)$$

Hence, through the design of $\mathbf{F_{RF}}\mathbf{F_{BB}}$, the precoder design optimization problem can be stated as,

$$(\mathbf{F_{RF}}^{opt}, \mathbf{F_{BB}}^{opt}) = \arg\max \mathbb{I}(\mathbf{F_{RF}}, \mathbf{F_{BB}})$$
$$\text{s.t. } \mathbf{F_{RF}} \in F_{RF}$$
$$||\mathbf{F_{RF}}\mathbf{F_{BB}}||_F^2 = N_s, \qquad (4.175)$$

where F_{RF} is the set of feasible RF precoders. The above problem is solved by first finding out the optimal precoder matrix $\mathbf{F}_{opt}$ from the SVD decomposition of the channel matrix $\mathbf{H}$ and taking the columns of the right unitary matrix corresponding to the largest N_s singular values. Then the precoders are designed by forming the following problem [EARAS+14].

$$(\mathbf{F_{RF}}^{opt}, \mathbf{F_{BB}}^{opt}) = \arg\min ||\mathbf{F}_{opt} - \mathbf{F_{RF}}\mathbf{F_{BB}}||_F,$$
$$\text{s.t. } \mathbf{F_{RF}} \in F_{RF}$$
$$||\mathbf{F_{RF}}\mathbf{F_{BB}}||_F^2 = N_s. \qquad (4.176)$$

In order to solve the above problem, an orthogonal matching pursuit algorithm similar to [TG07] is followed where the columns $\mathbf{F_{RF}}$ are gradually appended from an angle search space which consists of orthogonal phased array responses. The RF precoder columns are chosen by calculating the correlation of all the components of the angle space with the optimal precoder matrix and finding out the largest norm value. The baseband precoder is then designed following the least squares technique which gives a zero forcing solution.

5

Heterogeneous Networks

5.1 Femtocell-/Small Cell-based Heterogeneous Networks

Heterogeneous network (HetNet) architecture with low-power stations such as picocells, femtocells, and relays is one of the promising approaches to support enormous traffic demand and to provide ubiquitous quality of service (QoS) aware service. Particularly, the deployment of femtocells is actively considered by the mobile industry [KYF09], [3GP10], [tr208].

Picocells: Picocells consist of low-power base stations (BSs) typically to provide coverage in hot-spots such as stadium.

Femtocells: Femtocells consist of low-power base stations (BSs) typically to provide coverage in home or office.

5.1.1 Femtocells/Small cells

Femtocells or home eNodes (HeNBs) are the small BSs (also known as **small cells**) with low transmit power to be used to provide coverage for home or office users. The concept of femtocell implies using very low power BS placed indoor to provide access to a cell of few meter radius. The femtocell connects to the core network (CN) of the Telecom operator through a high data rate backhaul connectivity, which can be either X-DSL, fiber, etc., as shown in Figure 5.1.

- Advantage of deploying femtocells at home is that the same user equipment (UE) can be used for high data rate indoor connectivity as well as outdoor connection, whereas if wireless local area network (WLAN) and wireless metropolitan area network (WMAN) combination is used, then the UE would need two hardwares to support this. Call handover is also difficult in such a hybrid setup.
- It also reduces the operational expenditure (OPEX) to the operators by offloading traffic from the macrocell networks.

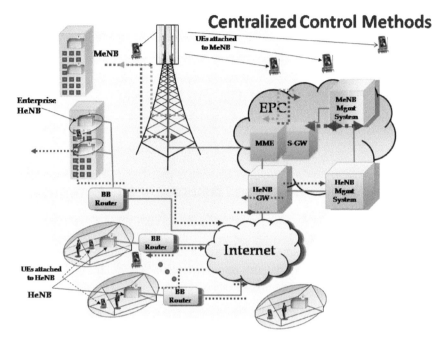

Figure 5.1   Illustration of femtocell network architecture.

- It is desired that when a user is indoor, calls are handled by the femtocells and channels from the macro/micro BS are freed. Thus, the system capacity is increased by offloading traffic from macro/micro BS to femtocells.
- The reason behind the interest toward small cell/femtocell deployment is motivated by the fact that 70% of high bit rate calls are originated from indoor [CAG08]. Since the indoor users with high data rate requirements consume a large portion of the available bandwidth, the number of supportable users in macrocell becomes less. Serving these users with high capacity demand through femtocells would help to reduce the bandwidth utilization in the macrocell, thereby increasing the number of supportable users in the macrocell which is known as the offloading gain.

Usually, the conventional BS deployment involves careful radio frequency (RF) planning considering coverage, interference issues, and QoS provisioning. As the femto BSs will be deployed by the users in a random and unplanned manner in large numbers, the co-channel interface (CCI) introduced by the femtocells must be controlled in order to ensure reliable services to both macrocell and femtocell users. Since the HetNet topology is different from conventional cellular networks, manual network deployment and maintenance is a tedious task. Therefore, the Third Generation Partnership Project (3GPP) standard has added several self-organizing network (SON) functionalities for femtocell network management [3GP08, 3GP12, 3gp05] in fourth-generation (4G) cellular systems. Further, modeling, analyzing, and designing of HetNets require a different approach than the conventional network dimensioning and management.

5.1.2 Deployment Modes

Femtocells can be operated in either dedicated mode or co-channel mode.

- **Co-channel Deployment:** In this mode of deployment, the femto BSs use the same spectrum used by the existing macro/micro BS. The system capacity can be increased several folds by deploying femtocells in co-channel mode due to full spatial reuse of total available spectrum compared to dedicated deployment mode [MG09]. However, it is achieved at the cost of degradation in the macrocell network performance due to increased CCI created by the randomly deployed femtocells. In single-frequency networks (SFNs) such as LTE and WiMAX, which are expected to provide enormous capacity, the CCI is the performance degradation factor that needs to be controlled carefully. Adding uncontrolled co-channel femtocells on top of these cellular systems would lead to further degradation in the performance of the systems.
- **Dedicated Channel Deployment:** In this mode of deployment, the femto BSs use different spectrum than used by the existing macro/micro BS.

5.1.3 Access Mechanisms

Open access and closed access are the two main access mechanisms being considered for femtocell deployment [GMP09].

- **Open Access:** Open access mode provides an improved overall system capacity compared to closed access mode as it allows any nearby user to access it. Also, the impact of CCI is less significant in open access mode.
- **Closed Access:** In closed access mode, since only the users registered to a particular femtocell are allowed to access it, the unregistered nearby

users suffer due to severe CCI. Since the access to the femtocells depends on the interest of owner of the femtocell and backhaul restrictions, closed access mode of deployment is under active consideration by the mobile industry [GMP09]. So, in this work, we focus on co-channel closed access femtocell deployment.

5.1.4 Issues Related to Femtocell/Small cell Deployment

The major issues that need to be solved for successful operation of femtocells within the existing macro/microcell networks are as follows:

- How femto BSs will adapt to its surrounding environment and allocate spectrum in the presence of intra- and cross-tier interference
- Timing and synchronization issues related to femtocell deployment
- Conflict avoidance for a mobile device under coverage of both the central BS and femto BS
- How will backhaul provide acceptable QoS?
- Whether to go for open access or closed access?
- How to handle Handoff in open access?
- What kind of billing/coverage policy should be used for roaming user going out of its own femto BS range?

5.1.5 Control Mechanisms for Femto Base Stations

The control mechanisms for femtocell deployment in a macro/micro cellular network can be classified as follows:

Distributed Setting: In closed access, controlling interference on data channels is an issue. Since a large number of femtocells are connected to the eNodeB Gateway (HeNB-GW), sharing the signaling information for all femto BSs may not be possible. Therefore, distributed auto-configuration is necessary for femto BSs to control CCI. The signal to interference plus noise ratio (SINR) and throughput performance of macro and femto UEs depends on the network load, the number of femtocells, and the location at which the femto BS is deployed inside the macrocell coverage area. Therefore, the femto BS has to change its transmit power and frequency assignment dynamically according to varying network conditions such as load condition at macrocell and femtocell networks.

Centralized Setting: In centralized setting, the central femto BS controller called HeNB-GW collects the networks statistics from both macrocell and femtocell networks such as macrocell network load, number of active femtocells, distribution of SINR of macrocell and femtocell UEs, time-of-day, etc. Based on the network statistics, the central controller estimates the combination of transmit power and load that maximizes the total system throughput while satisfying QoS requirements of macrocell and femtocell networks. Then the femtocell management system sends information to the femto BSs on appropriate transmit power levels and load through broadcast control channel (BCCH) or backhaul. This information passing to the femtocells could be in hourly basis.

Hybrid Setting: The central HeNB controller sends an upper limit on allowable transmit power, bandwidth, and throughput for the femtocell. Then the femtocell dynamically adjusts the power and load over and above the allowed limit within a certain margin to meet the rate requirement. Then the central controller may also inform the management entity in the CN to adjust the active number of UEs and load in the macro/micro cell depending upon the objectives.

5.1.6 Area Spectral Efficiency Analysis of Co-Channel Heterogeneous network

5.1.6.1 Spectral efficiency of a macrocell network
The SINR for a macrocell user, denoted by $\gamma_m(\chi, h, v)$, is a function of interferer's activity (v), shadow fading component (χ), and small-scale fading component (h). The received SINR for a macrocell user at position $\dot{r}_m = (r_m, \theta_m)$ on subchannel k can be expressed as

$$\gamma_m^k(\chi, h, v) = \frac{P_{tkm_0} G_{km_0}}{\sum\limits_{i=1}^{N_m-1} P_{tkm_i} v_{km_i} G_{km_i} + \sum\limits_{j=1}^{N_{fint}} P_{tkf_j} v_{kf_j} G_{kf_j} + B_{sb} N_0}.$$

(5.1)

where P_{tkm_0} is the transmit power of serving macro BS on subchannel k, P_{km_i} and P_{kf_j} are the transmit powers of i-th macro BS and j-th femto BS respectively on subchannel k. The Gamma distributed channel powers between the user and i-th macro BS and j-th femto BS on subchannel k are z_{km_i} and z_{kf_j} respectively. The activity status of the i-th macro BS and j-th femto BS on subchannel k are denoted by v_{km_i} and v_{kf_j} respectively, and N_0 is noise power in (Watts/Hz).

The probability distribution function (PDF) of SINR $\gamma_m^k(\chi, h, v)$ at position $\dot{r}_m$ is given by

$$p_{\gamma_m}(x|\dot{r}_m) = \frac{1}{x\sigma_{\gamma_m}(r_m, \theta_m)\sqrt{2\pi}} \exp\left[\frac{-(\ln(x) - \mu_{\gamma_m}(r_m, \theta_m))^2}{2\sigma_{\gamma_m}^2(r_m, \theta_m)}\right].$$

$$(5.2)$$

The above PDF expression is conditioned on the location of the user. By integrating the above expression over r_m and θ_m, the area averaged PDF of SINR is obtained as

$$p_{\gamma_m}(x) = \int_{R_{m0}}^{R_c} \int_0^{2\pi} p_{\gamma_m}(x|\dot{r}_m)\, p_{\dot{r}_m}(r_m, \theta_m)\, dr\, d\theta$$

$$= \frac{1}{(R_c - R_{m0})^2\sqrt{2\pi^3}}$$

$$\times \int_{R_{m0}}^{R_c} \int_0^{2\pi} \frac{(r_m - R_{m0}) \exp\left[\frac{-(\ln(x) - \mu_{\gamma_m}(r_m, \theta_m))^2}{2\sigma_{\gamma_m}^2(r_m, \theta_m)}\right]}{x\sigma_{\gamma_m}(r_m, \theta_m)} dr_m\, d\theta_m.$$

$$(5.3)$$

Since there is no closed-form expression available for the above integral, the PDF is obtained using numerical integration. Note that the location-dependent mean $\mu_{\gamma_m}(r_m, \theta_m)$ and standard deviation $\sigma_{\gamma_m}(r_m, \theta_m)$ in (5.3) are the function of femtocell parameters: transmit power (P_{tf}), load (β_f), and density (ρ_f) as well as macrocell parameters: BS transmit power (P_{tm}) and load condition (β_m). Therefore, the average spectral efficiency (SE) obtained from (5.3) will also be the function of these parameters. Finally, the mean SE of the macrocell in each subchannel, (C_m) in (b/s/Hz), is obtained using (5.4) below,

$$C_m = B \times \sum_{l=1}^{N_L} p_l^{mcs} b_l \quad \text{(b/s)}, \qquad (5.4)$$

where N_L is the number of modulation and coding scheme (MCS) levels, p_l^{mcs} is the probability of utilizing the l-th MCS which is obtained from (5.3), and b_l is the number of bits transmitted using the l-th MCS level.

5.1.6.2 Spectral efficiency of a femtocell network
Consider a femtocell of radius R_f located at distance d from the macro BS. The received SINR on subchannel k for a femtocell user located inside the

femtocell at position $\dot{r}_f = (r_f, \theta_f)$ can be expressed as

$$\gamma_f^k(\chi, h, v) = \frac{P_{tkf_0} G_{kf_0}}{\sum\limits_{i=1}^{N_m} P_{tkm_i} v_{km_i} G_{km_i} + \sum\limits_{j=1}^{N_{fint}} P_{tkf_j} v_{kf_j} G_{kf_j} + B_{sb} N_0}. \quad (5.5)$$

Assume that the femtocell user positions are uniformly distributed within the circular area with radius R_f. The area averaged PDF of SINR $\gamma_f^k(\chi, h, v)$ of the femtocell is obtained by

$$p_{\gamma_f}(x) = \frac{1}{(R_f - R_{f0})^2 \sqrt{2\pi^3}}$$

$$\times \int_{R_{f0}}^{R_f} \int_0^{2\pi} \frac{(r_f - R_{f0}) \exp\left[\frac{-(\ln(x) - \mu_{\gamma_f}(r_f, \theta_f))^2}{2\sigma_{\gamma_f}^2(r_f, \theta_f)}\right]}{x \sigma_{\gamma_f}(r_f, \theta_f)} dr_f \, d\theta_f,$$

$$(5.6)$$

where R_{f0} is the minimum distance between the user and femto BS, and $\mu_{\gamma_f}(r_f, \theta_f)$ and $\sigma_{\gamma_f}(r_f, \theta_f)$ are the mean and standard deviation of the SINR of the femtocell user at location $\dot{r}_f$ from the femto BS which are obtained in the same way as $\mu_{\gamma_m}(r_m, \theta_m)$ and $\sigma_{\gamma_m}(r_m, \theta_m)$ are obtained in the macrocell PDF of SINR expression.

The range of distances from the central macro BS over which the SINR distribution of femtocells is the same can be considered as a ring. Therefore, we divide the circular disc with radius R_c into Q annular rings with inner radius R_q and outer radius R_{q+1} as described in [CD14]. The average SE achieved by the femtocells in the q-th ring C_{f_q} is obtained from the SINR distributions of femtocells in q-th ring by using (5.4) and (5.6). Since it is assumed that the femtocells are uniformly distributed within the macrocell, the number of femtocells located within the q-th annular ring can be written as $N_{F_q} = \frac{\pi(R_{q+1}^2 - R_q^2)}{|\mathcal{A}|} = \frac{2\pi(R_{q+1}^2 - R_q^2)N_f}{3\sqrt{3}N_f Rc^2}$. Finally, the total SE achieved by all the femtocells distributed within the central macrocell is obtained as

$$C_f = \sum_{q=1}^Q C_{f_q}^{(tot)} = \sum_{q=1}^Q \frac{R_{q+1}^2 - R_q^2}{R_c^2} N_f \, C_{f_q}. \quad (5.7)$$

Note that here $R_1 >$ the minimum distance between the macro BS and femto BS R_{mf0}.

5.1.6.3 Area spectral efficiency

Let $C_m(\beta_m, \beta_f, P_{tm}, P_{tf}, \rho_f)$ and $C_f(\beta_m, \beta_f, P_{tm}, P_{tf}, \rho_f)$ be the SEs (in b/s/Hz) achieved by the macrocell and femtocell network, respectively, in each subchannel with N_f number of femtocells at macrocell load, β_m, and femto load, β_f, for femtocell transmit power, P_{tf}, and macrocell transmit power, P_{tm}. Let $T_{um,min}$ and $T_{uf,min}$ be the minimum required throughputs of a macrocell user and a femtocell user, respectively. The average cell throughput (b/s) achieved by the macrocell at load condition β_m can be obtained as $T_m(\beta_m, \beta_f, P_{tm}, P_{tf}, \rho_f) = \beta_m B C_m(\beta_m, \beta_f, P_{tm}, P_{tf}, \rho_f)$.

The long-term average throughput achieved by the macrocell user is $\overline{T}_{um}(\beta_m, \beta_f, P_{tm}, P_{tf}, \rho_f) = \frac{T_m}{N_{um}^{act}(\beta_m)}$, where $N_{um}^{act}(\beta_m)$ is the number of users served by the macro BS at load β_m. The total throughput achieved by the femtocell network at load condition β_m can be obtained as $T_f(\beta_m, \beta_f, P_{tm}, P_{tf}, \rho_f) = \beta_f B C_f(\beta_m, \beta_f, P_{tm}, P_{tf}, \rho_f)$. The total number of femtocell users can be written as $N_{uf}^{tot} = \sum_{j=1}^{N_f} N_{uf_j}$. Here, N_{uf_j} is the number of users supported by the j-th femtocell which is given by $N_{uf_j} = \frac{\beta_f B C_{f_j}}{T_{uf,min}}$, where C_{f_j} is the long-term SE (in b/s/Hz) achieved by the j-th femtocell. The long-term average throughput achieved by the femtocell user is $\overline{T}_{uf}(\beta_m, \beta_f, P_{tf}, \rho_f) = \frac{T_f}{T_{uf}^{tot}}$. For a given network condition, the total throughput achieved by the macrocell and all the femtocells located within its coverage area, $|\mathcal{A}|$, can be written as

$$
\begin{aligned}
T_T(\beta_m, \beta_f, P_{tm}, P_{tf}, \rho_f) &= T_m(\beta_m, \beta_f, P_{tm}, P_{tf}, \rho_f) \\
&\quad + T_f(\beta_m, \beta_f, P_{tm}, P_{tf}, \rho_f) \quad \text{(b/s)} \\
&= \beta_m B C_m(\beta_m, \beta_f, P_{tm}, P_{tf}, \rho_f) \\
&\quad + \beta_f B C_f(\beta_m, \beta_f, P_{tm}, P_{tf}, \rho_f).
\end{aligned}
\tag{5.8}
$$

The total area spectral efficiency (ASE) achieved by the macrocell-femtocell network can be expressed as

$$
ASE_T(\beta_m, \beta_f, P_{tm}, P_{tf}, \rho_f) = \frac{T_T(\beta_m, \beta_f, P_{tm}, P_{tf}, \rho_f)}{B.|\mathcal{A}|} \quad \text{(b/s/Hz/m}^2\text{)}
\tag{5.9}
$$

5.1.6.4 Optimal femtocell radio parameters

In order to achieve the benefit of femtocell deployment, i.e., to achieve maximum ASE, the CCI caused by the femtocell network must be controlled so that the QoS conditions of both macrocell and femtocell networks remain unaffected.

Since the interference situation depends on the load condition in the macrocell network as well in the femtocell network and number of active femtocells, the femtocells should adapt their radio parameters according to variation in the network condition. The CCI can be kept within the tolerable level by appropriately controlling either the transmit power (P_{tf}) or the load (β_f) of femtocells. We assume that the femto BSs allocate a set of random subchannels to its users from the total available subchannels. This provides randomized interference avoidance, because the random subchannel allocation would result in less probability of subchannel collisions between the macrocell and femtocell networks [CA09, GI10].

Hence, for a given network condition, macrocell load condition ($\beta_m = \mathcal{B}_m$), macrocell transmit power ($P_{tm} = \mathcal{P}_{tm}$), number of femtocells ($N_f = \mathcal{N}_f$), and the optimal combination of femtocell transmit power and load that maximizes the ASE while satisfying QoS constraints in both the networks can be obtained from the following optimization problem,

$$\{P_{tf}^*, \beta_f^*\} = \underset{\{P_{tf}, \beta_f\}}{\arg\max} \ ASE_T(\mathcal{B}_m, \beta_f, \mathcal{P}_{tm}, P_{tf}, \mathcal{N}_f), \qquad (5.10)$$

$$\text{subject to:} \quad C1 : \overline{T}_{um}(\mathcal{B}_m, \beta_f, \mathcal{P}_{tm}, P_{tf}, \mathcal{N}_f) \geq T_{um,min}$$
$$C2 : \overline{T}_{uf}(\mathcal{B}_m, \beta_f, \mathcal{P}_{tm}, P_{tf}, \mathcal{N}_f) \geq T_{uf,min}$$
$$C3 : P_{tf,min} \leq P_{tf} \leq P_{tf,max}.$$
$$C4 : 0 \leq \beta_f \leq 1.$$

Here the constraints $C1$ and $C2$ denote the QoS requirement of macrocell and femtocell users, respectively. Constraints $C3$ and $C4$ represent the range of optimization variables (femtocell transmit power and load), respectively. Since we derive the total ASE, ASE_T semi-analytically from the SINR distributions of macrocells and femtocells, it cannot be explicitly described as a function of the optimization variables P_{tf} and β_f. However, the problem can be described using *oracle* model (see Section 4.1.4 in [BV04]). The *oracle* model is used when the objective function $f(x)$ is not known explicitly, but the function and its derivatives $f'(x)$ (using finite difference method) can be evaluated at any x whenever they are required during the optimization process.

The ASE gain, G_{ASE} is defined as the ratio of the ASE achieved by the macrocell–femtocell networks to the ASE achieved by the macrocell with no femtocell deployment.

$$G_{ASE} = \frac{T_m(\beta_m, \beta_f, P_{tm}, P_{tf}, \rho_f) + T_f(\beta_m, \beta_f, P_{tm}, P_{tf}, \rho_f)}{T_m(\beta_m, \beta_f, P_{tm}, P_{tf}, \rho_f = 0)}. \quad (5.11)$$

5.1.6.5 Results and discussion

The results presented in this section are based on the system parameters as shown in Table 5.1. It is assumed that the macrocells are uniformly loaded, i.e., the load factor in all the macrocells in the network is the same. The same transmit power and load is assumed for all the femtocells located within the macrocell coverage area. This section presents the results to show the impact of femtocell's and radio access network's radio parameters on the ASE and average macrocell user throughput.

Figure 5.2(a) shows the impact of varying femtocell transmit power (P_{tf}) and load (β_f) on the total macrocell throughput with 50 femtocells. From the figure, it can be seen that operating femtocells with high transmit power and load severely affects the macrocell throughput performance. Figure 5.2(b)

Table 5.1   System parameters used for ASE analysis of macrocell–femtocell networks

Parameter	Value
Number of macrocells (N_m)	19
Macrocell/femtocell radius (R_c, R_f)	288 m, 20 m
Carrier frequency (f_c)	2.4 GHz
Macrocell transmit power (P_{tm})	43 dBm
Minimum distance between the user and macro BS (R_{m0})	10 m
Minimum distance between the user and femto BS (R_{f0})	1 m
Minimum distance between the femto and macro BSs (R_{mf0})	20 m
Minimum and maximum femtocell transmit power ($P_{tf,min}, P_{tf,max}$)	−10 dBm, 20 dBm
Minimum required macrocell user throughput ($T_{um,min}$)	128 kbps
Minimum required femtocell user throughput ($T_{uf,min}$)	5 Mbps
Bandwidth (B)	10 MHz
Thermal noise level (N_0)	−174 dBm/Hz
Number of MCS levels (N_L)	8
Subchannel bandwidth (B_{sb})	200 KHz
Pathloss exponents (α_m, α_f)	4, 3
Shadow parameters ($\sigma_{\xi_m}, \sigma_{\xi_f}$)	6 dB, 4 dB
Wall penetration loss (L_w)	10 dB

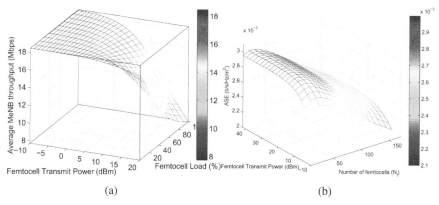

Figure 5.2 (a) Average macrocell throughput for varying femtocell Tx. power and load. (b) ASE for varying femtocell Tx. power and number of femtocells.

shows the impact of increasing femtocell transmit power and femtocell density on the ASE ($\frac{T_M + T_F}{|A_m|}$) performance. With low femtocell density, increasing the transmit power of femtocells significantly increases the ASE due to increased femtocell network throughput, whereas increasing number of femtocells results in reduced poor ASE performance. This is because, due to a large number of femtocells, severe interference caused by neighboring femtocells reduces the throughput achieved by the femtocell network.

5.1.6.6 ASE with QoS constraints

This subsection presents the results of maximum achievable ASE (ASE_T^*) with the optimal femtocell transmit power P_{tf}^* which is obtained from with fixed femtocell load ($\beta_f = \mathcal{B}_f$) for different macro load conditions. The required average throughput for macrocell users, $T_{um,min}$ is set to 128 kbps and for femtocell users, $T_{uf,min}$: 5 Mbps. The minimum and the maximum femtocell transmit power levels considered for femto BSs is −10 and 20 dBm, respectively [www10].

Figure 5.3 shows the femtocell transmit power and load values obtained from (5.10) at different macrocell load conditions for 80 femtocells with P_{tm} = 43 dBm. It is seen that that the lower femtocell transmit power is needed for higher femtocell load in order to satisfy QoS requirements of macro and femto UEs, whereas when femtocell load is low, higher transmit power is sustainable. Further, the allowable femtocell transmit power level is much higher at low macro load condition compared to full load condition.

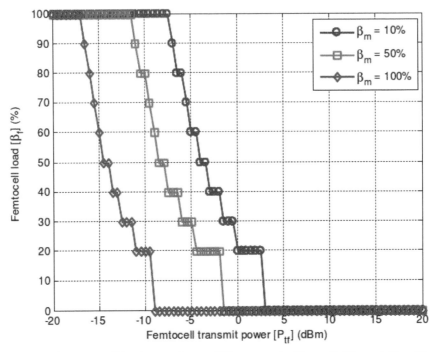

Figure 5.3   Femtocell load versus femtocell Tx. power, for P_{tm} = 43 dBm and N_f = 80.

Figure 5.4(a) shows the ASE achieved with optimal transmit power levels for different femto and macro cell load conditions. The corresponding number of femtocells per cell site is shown in Figure 5.4(b). It can be found that for high macro load, and high femtocell load, the ASE and number of femtocells increase up to a certain point and then decrease with any further increase in femtocell load. This is because femtocell load causes heavy CCI to the nearby femtocells due to high activity of femtocells which results in failing to satisfy the QoS condition in a femtocell network. Thus, supportable femtocell density becomes limited which results in reduced ASE. It can also be observed that for a given macro load condition, the maximum ASE and the highest number of supportable femtocells occurs at different femtocell load values. This brings out the fact that there is a tradeoff between the ASE and the number of supportable femtocells.

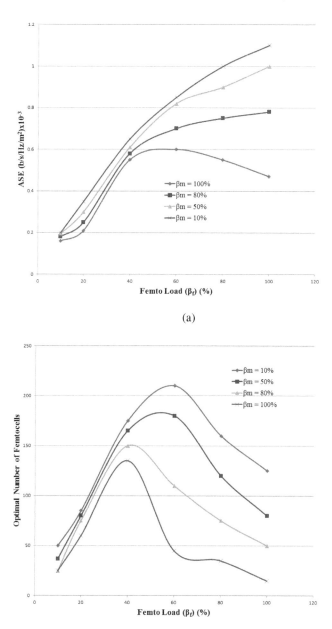

Figure 5.4 (a) Mean ASE versus Femtocell load with optimal femtocell transmit power. (b) Maximum number of allowable femtocells per cell site versus femtocell load with optimal femtocell transmit power.

5.2 OFDMA-based Cellular Network and Underlaying D2Ds

This part presents the performance of D2D underlaying the OFDMA-based cellular network.

Device to device (D2D) communication is a promising concept that helps to fulfill bandwidth hungry applications and services, leading to better utilization of radio resources, shorter delay, better user experience, better frequency reuse, and public safety use cases. D2D communications enable two proximal UEs/devices to directly communicate with each other, thereby bypassing the BS. One of the most important perspectives of D2D would be the energy saving as two proximal D2D devices would communicate with less transmitter power. It may operate in either cellular controlled and in licensed band or self-controlled and in unlicensed band. Offloading from cellular networks, improving the battery lifetime of UEs, and creation of new services are the other advantages associated with D2D. Use cases of D2D could be cooperative communications, social networking, social commerce and advertisements, augmented reality services and broadcasting personal data/info, etc. [HH15, KBM+15, MSS+14, KLNK14, VF15, SWSQ15].

3GPP has proposed D2D communication, as one of the promising techniques to enhance the performance of LTE-A. D2D communication enables improvement in terms of overall system capacity, number of UEs in connected state, and evolved node-B (eNB) offloading gains. D2D communication is an *ad hoc* process where two proximal UEs communicate directly instead of communicating through eNB [DRW+09] but the D2D process is controlled by eNB. The major functions of D2D communications include

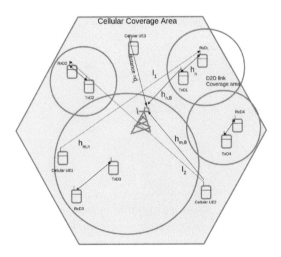

peer discovery, physical layer procedure, and radio resource management (RRM). Mode selection, resource allocation, power control, and inter-cell and intra-cell interference mitigation are RRM issues in D2D communications [FDM⁺12]. Mode selection deals with whether to serve a UE in cellular mode or D2D mode. The eNB plays a key role in mode selection and it depends on the proximity of devices, inter-cell and intra-cell interference conditions, channel conditions, and instantaneous load condition on the network.

5.2.1 D2D Deployment Modes

D2Ds can be operated in either dedicated mode or co-channel mode.

- Co-channel Deployment: In this mode of deployment, the D2Ds operate using the same spectrum as used by macro/micro BSs of cellular system. Due to spatial reuse of spectrum, there is an increase in system capacity but at the cost of higher CCI. In SFN such as LTE, which operates in the interference limited region, increase in CCI due to D2D communication poses further challenge to the system design.
- Dedicated Channel Deployment: In this mode of deployment, the D2D and macro BSs use different frequency spectra in the total available bandwidth. In this mode, both the D2D and macro users do not experience cross-tier interference as both the macrocells and D2Ds use different frequency channels; however, the spectrum may not be as efficiently used as in the previous case.

5.2.2 System Level Performance Evaluation

The system performance can be evaluated using analytical deterministic geometry-based models as well as using stochastic geometry approaches. As the current cellular networks are becoming HetNet due to the simultaneous presence of remote radio heads, micro BSs, pico BSs, femtos, relay nodes, etc., and because of randomness in the cell size, coverage, and random placement of these nodes, the evaluation method best suited is via stochastic geometry approach [DGBA12b].

5.2.3 Stochastic Modeling and Analysis of D2D Enabled Heterogeneous Cellular Network

ζ-Ginibre point process (GPP) is considered for BS distribution in heterogeneous cellular networks (HCNs) to deal with the superimposition flaw

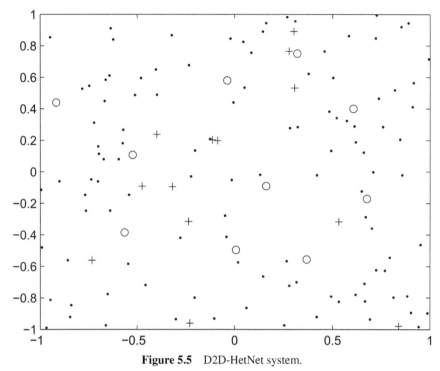

Figure 5.5 D2D-HetNet system.

associated with Poisson point process (PPP) and to incorporate the repulsive nature. This chapter considers ζ- GPP HCN where micro BSs (eNBs) are ζ-GPP distributed and femtos and D2D communications are PPP distributed.

- We consider a downlink single-frequency OFDMA-based multi-cellular system. The microcell BSs in the network are distributed by ζ-GPP (small circles in Figure 5.5). Each cell has system bandwidth B_{sys}.
- The UEs (D2D transmitters and receivers) and femtos are PPP distributed (dots and plus sign, respectively, in Figure 5.5).
- Downlink resources are used for cellular, D2D, and for femto tier as well.

5.2.3.1 Distribution of eNBs

The distribution of eNBs is taken to be ζ-GPP distributed and is described as follows. Let ϕ be a realization of a certain point process with observation window W. The n-th joint density function of a point process on Lebesgue measure $\mathbb{R}^2$ with Borel set $B \subset R^d$ is defined by

$$\mathbb{E}\left[\prod_{i=1}^{n} \phi(B_i)\right] = \int_{B_1 \times B_2 \ldots \times B_n} w^{(n)}(x_1, .., x_n)dx_1...dx_n \qquad (5.12)$$

where w^1 is the spatial density of nodes (eNBs). In the above, $w^{(n)}(x_1, .., x_n)dx_1...dx_n$ represents the probability of locating a node in the vicinity of x_i. The ζ-GPP is a determinantal point process that can be defined by map $K_{c,\zeta}$ called as kernel which represents the interaction force between the different points of the point process and can be expressed as

$$w^{(n)}(x_1, ..., x_n) = \det(K_{c,\zeta}(x_i, x_j)), 1 \leq i, j \leq n), \qquad (5.13)$$

where ζ be a real number $]0,1]$ and $c = \lambda\pi$ and here λ be strictly positive real number be the intensity of the point process. Then kernel can be expressed as in [DZH15, KP14, KM14] as

$$K_{c,\zeta}(x, y) = \frac{c}{\pi}e^{-\frac{c}{2\zeta}(|x|^2+|y|^2)}e^{\frac{c}{\zeta}x\bar{y}} \qquad (5.14)$$

x,y $\in B(0, R)$ and

$$K(r) = \pi r^2 - (\frac{\zeta\pi}{c}(1 - e^{-\frac{c}{\zeta}r^2}). \qquad (5.15)$$

whereas for homogeneous PPP, $K(r)$ is πr^2.

For non-Poisson distribution, an important term is the J function in terms of location r and for ζ-GPP which is given by

$$J(r) = \frac{1 - G(r)}{1 - F(r)} = (1 - \zeta + \zeta e^{-\frac{c}{\zeta}r^2})^{-1} \qquad (5.16)$$

for $r \geq 1$ and where G is the nearest neighbor function and F is the empty space function. In case of uniform PPP with intensity λ, $F(r) = G(r) = 1 - e^{-\lambda\pi r^2}$ and J(r) = 1. The properties of J(r) are, if

$$J(r) = 1, \text{Poisson distribution}$$
$$J(r) < 1, \text{Custered distribution}$$
$$J(r) > 1, \text{Regular distribution}$$

This functional gives idea/test of the strength of interpoint interaction. In clustered distribution, each parent point "a" gives rise to "b" offspring points according to certain stochastic mechanism, e.g., Matern cluster and Thomas cluster process. In clustered distribution, points are clumped whereas in regular distribution, points are spaced due to certain stochastic mechanism, e.g., ζ-GPP.

5.2.3.2 Distribution of UEs (D2D Tx and Rx) and femtos

Distributions of D2D transmitters (UEs) as well as of femtos are considered to follow PPP distribution. For a given λ_d and when D2D transmitters are uniformly distributed PDF of distance r of the n-th nearest D2D transmitter from the reference D2D transmitter can be expressed as

$$f_{r_n}(r, n) = \frac{2(\pi\lambda_d)^n}{(n-1)!} r^{2n-1} e^{-\pi\lambda_d r^2}, \ r \geq 0, \ n = 1, 2, \dots. \tag{5.17}$$

For the given λ_d and PDF, the expected distance of the n-th neighboring D2D transmitter can be given by

$$E[r_n] = \bar{r}_n = \int_0^\infty r \, f_{r_n}(r, n) \, dr = \frac{\Gamma\left(n + \frac{1}{2}\right)}{\sqrt{\pi\lambda_d}(n-1)!}, \tag{5.18}$$

where $\Gamma(x) = \int_0^\infty e^{-t} t^{x-1} dt$ and for femto as $E[r_n] = \bar{r}_n = \frac{\Gamma\left(n + \frac{1}{2}\right)}{\sqrt{\pi\lambda_f}(n-1)!}$.

5.2.4 Signal-to-interference Ratio with Activity

The reference UE will receive the signal from the intended eNB in cellular tier or intended D2D transmitter in D2D tier or from intended femto in femto tier. All modes would be affected by interference from other D2D transmitters and the transmission from other eNBs in the neighboring cells as well as from femtos too. The signal-to-interference ratio (SIR) experienced by UE at an arbitrary distance $(r_{i,m})$ from the ith transmitter of the $m^{\text{th}} \in \{c, d, f\}$ tier on a sub-band k can be expressed as

$$\gamma_k(r_{i,m}) = \frac{P_{i,m}|h_{i,m}|^2 r_{i,m}^{-\eta}}{\sum_{j \in \phi_c} P_c |h_{j,c}|^2 r_{j,c}^{-\eta} + \sum_{u \in \phi_d} v_d P_d |h_{u,d}|^2 r_{u,d}^{-\eta} + \sum_{l \in \phi_f} v_f P_f |h_{l,f}|^2 r_{l,f}^{-\eta}} \tag{5.19}$$

where the subscript $m \in \{c, d, f\}$ indicates tier, c denotes cellular tier, d denotes D2D tier, and f denotes femto tier. In the numerator, $P_{i,m}$ is the power transmitted from the intended BS and $r_{i,m}^{-\eta}$ is the path loss component, whereas $|h_{i,m}|^2$ is the gain due to small-scale fading (follows exponential distribution). In the denominator, the first, second, and third terms are interference due to other eNBs, D2D transmitters, and femtos, respectively. Here v_d and v_f are the activity factors associated with D2D and femto tier, respectively. Here a particular tier all BSs have the same transmitted power. The outage probability may be expressed as

$$1 - p_{cov} = \mathbb{E}\left(\prod_{q=1}^{M} \prod_{r_i \in \phi_q} 1(\gamma_k(r_{i,m}) < \gamma_{th,q}) \right) \qquad (5.20)$$

Expanding the inner product and as higher order terms become zero, we get as in [DGBA12b] as

$$p_{cov} = \sum_{q=1}^{M} \mathbb{E} \sum_{r_i \in \phi_q} 1(\gamma_k(r_{i,m}) > \gamma_{th,q}) \qquad (5.21)$$

using Campbell Mecke theorem, $p_{cov} = \sum_{q=1}^{M} \frac{\lambda_q}{N_{sb}} \int_{R^2} \mathbb{P}(\gamma_k(r_{i,m}) > \gamma_{th,q}) dr$

5.2.5 Sub-band Coverage Probability Under PPP

The sub-band coverage probability of a UE (in subband k) associated to a BS with maximum link SIR among all other BSs (considering PPP distributed nodes in all tiers), can be expressed as in [JSXA12] as

$$p_{cov} = \mathbb{P}(\gamma_k(r_{i=1,m}) > \gamma_{th})$$
$$= \sum_{m=1}^{M} \frac{\lambda_m}{N_{sb}} \int_0^\infty \exp\left(-r^2 Q(\eta)(\frac{\gamma_{th}}{P_m})^{2/\eta} \sum_{q=1}^{M} \frac{\lambda_q}{N_{sb}} (P_q)^{2/\eta} \right) dr, \qquad (5.22)$$

where $Q(\eta) = \frac{2\pi^2}{\eta} \csc(\frac{2\pi}{\eta})$, where $Q(\eta) = \frac{2\pi^2}{\eta} \csc(\frac{2\pi}{\eta})$

5.2.6 Coverage Probability Under ζ GPP

$$p_{cov} = \mathbb{P}(\gamma > \gamma_{th}) = P(\gamma > \gamma_{th}, B_0 = i) = \mathbb{P}(|h_{i,m}|^2$$
$$> \frac{\gamma_{th}}{P_i^{m,q} r_{i,m}^{-\eta}} \sum_{j \in \phi_c} P_c |h_{j,c}|^2 r_j^{-\eta}, |r_i| < |r_j|) \qquad (5.23)$$

as $|h_i|^2$ follows exponential distribution with unit mean. Also ζ Ginibre point process ϕ is with intensity λ/ζ with independent marks $P(\xi_i = 1) = \zeta$ and $P(\xi_i = 0) = 1 - \zeta$ with $I = P_c \sum_{j \in \phi_c} |h_i^{c,q}|^2 r_i^{-\eta} = P_c \sum_{j \in \phi_c | i} \xi_j |h_{j,c}|^2 r_{jc,q}^{-\eta}$; the above can be expressed as in [MS16, DZH15] as

$$\mathbb{P}(\gamma > \gamma_{th}) = P(\xi_i = 1) E\left[e^{-\frac{\gamma_{th} I}{P_{i,m} r_{i,m}^{-\eta}}}, A_i \right] = E\left[\zeta E\left[e^{-\frac{\gamma_{th} I}{P_{i,m} r_1^{-\eta}}} \right] 1_{A_i} \right]$$
(5.24)

where A_i is the area under ith transmitter.

Further, $|h_{j,c}|^2$ also follows exponential distribution with unit mean; the above can be expressed as

$$p_{cov} = E\left[\zeta \prod_{j \in \phi_c | i=1} \left[1 - \zeta + \frac{\zeta}{1 + \gamma_{th}(\frac{r_j}{r_1})^{-\eta}} 1_{\bar{r}_j > \bar{r}_i} \right] \right]$$
(5.25)

5.2.7 User Association under Cellular Tier Being ζ GPP and D2D and Femto Being PPP Distributed

User association under the BS with strongest power criteria can be illustrated for D2D enabled HCN (three, tier network) in the following lines.

As in [WDZH16] and [GH16], approximate (ζ GPP) non-PPP coverage probability can be expressed in terms of PPP by scaling the threshold value. Thus, as in [WDZH16] and [DGBA12a] under the strongest BS association it can be given as

$$\mathbb{P}(\gamma(r_{i=1,m}) > \gamma_{th})$$
$$\approx \int_0^\infty \exp\left(-r\left(\frac{\lambda_c}{N_{sb}}\right)^{-1} P_c^{-2/\eta} Q(\eta) \left[\frac{\lambda_c}{N_{sb}} \left(P_c \frac{\gamma_{th}}{1 + \zeta/2} \right)^{2/\eta} \right. \right.$$
$$\left. \left. + \frac{\lambda_d}{N_{sb}} (P_d \gamma_{th})^{2/\eta} + \frac{\lambda_d}{N_{sb}} (P_f \gamma_{th})^{2/\eta} \right] \right) dr,$$
(5.26)

D2D tier as

$$\mathbb{P}(\gamma(r_{i=1,m}) > \gamma_{th})$$
$$\approx \int_0^\infty \exp\left(-r\left(\frac{\lambda_d}{N_{sb}}\right)^{-1} P_d^{-2/\eta} Q(\eta) \left[\frac{\lambda_c}{N_{sb}} \left(P_c \frac{\gamma_{th}}{1 + \zeta/2} \right)^{2/\eta} \right. \right.$$
$$\left. \left. + \frac{\lambda_d}{N_{sb}} (P_d \gamma_{th})^{2/\eta} + \frac{\lambda_d}{N_{sb}} (P_f \gamma_{th})^{2/\eta} \right] \right) dr,$$
(5.27)

Femto tier as

$$
\mathbb{P}(\gamma(r_{i=1,m}) > \gamma_{th})
$$

$$
\approx \int_0^\infty \exp\left(-r\left(\frac{\lambda_f}{N_{sb}}\right)^{-1} P_f^{-2/\eta} Q(\eta) \left[\frac{\lambda_c}{N_{sb}}\left(P_c \frac{\gamma_{th}}{1+\zeta/2}\right)^{2/\eta}\right.\right.
$$

$$
\left.\left. + \frac{\lambda_d}{N_{sb}}(P_d\gamma_{th})^{2/\eta} + \frac{\lambda_d}{N_{sb}}(P_f\gamma_{th})^{2/\eta}\right]\right)dr. \tag{5.28}
$$

5.2.8 Fractional Load and Link Throughput

The average load on the mth tier over sub-band k can be expressed as in [DGBA12a] as

$$
\bar{\beta}_{k,m} = \frac{\frac{\lambda_m}{N_{sb}}(P_m\gamma_{th})^{2/\eta}}{\left[\frac{\lambda_c}{N_{sb}}\left(P_c\frac{\gamma_{th}}{1+\zeta/2}\right)^{2/\eta} + v_d\frac{\lambda_d}{N_{sb}}(P_d\gamma_{th})^{2/\eta} + v_f\frac{\lambda_f}{N_{sb}}(P_f\gamma_{th})^{2/\eta}\right]} \tag{5.29}
$$

Probabilistic link throughput over sub-band k with bandwidth B_{sb} using round robin at a distance r_i from a random transmitter belonging to the mth tier under load β_m may be expressed as

$$
T_{k,m}(r_i) = B_{sys}\frac{\frac{\lambda_m}{N_{sb}}(P_m\gamma_{th})^{2/\eta}}{\left[\frac{\lambda_c}{N_{sb}}\left(P_c\frac{\gamma_{th}}{1+\zeta/2}\right)^{2/\eta} + v_d\frac{\lambda_d}{N_{sb}}(P_d\gamma_{th})^{2/\eta} + v_f\frac{\lambda_f}{N_{sb}}(P_f\gamma_{th})^{2/\eta}\right]}
$$

$$
\times \log_2\left[1 + \gamma_{th}(r)\right]\mathbb{P}\left[\gamma_k(r_{i,m}) > \gamma_{th}\right] \tag{5.30}
$$

5.2.9 Results and Their Analysis

The parameters used are given in Table 5.2 for analytical and simulation analysis.

5.2.10 Coverage Probability Analysis

Figure 5.6 shows the coverage probability versus γ_{th} plot for ζ-GPP when the transmitter is placed 25 m from the receiver for cellular, D2D, and femto. Here D2D activity factor and femto activity factor = 50%. Solid lines indicate coverage probability for analytical whereas dashed lines indicate coverage probability obtained through simulation.

It is seen that the coverage probability decreases with the increase in γ_{th}. It is also seen that the p_{cov} for ζ -GPP cellular is always higher for all γ_{th}

Table 5.2 System parameters

(Parameter)	(Value)
P_c	41 dBm
P_d	24 dBm
P_f	20 dBm
Simulation area	$\pi 288^2$
λ_c	1/Simulation area
λ_d	50/Simulation area
λ_f	10/Simulation area
Path loss exponent	3
Bandwidth (B)	10 MHz

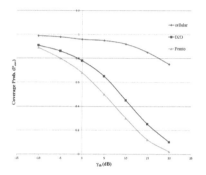

Figure 5.6 Coverage probability versus threshold, for 25 m.

as compared to D2D and femto. Further observe that D2D p_{cov} is better than that of femto.

5.2.11 Throughput Analysis

Figure 5.7(a) shows the cellular, D2D, and femto throughput versus transmitter distance from the receiver with D2D and femto activity factor equal to 25% and 12%, respectively, at a cellular load $\beta_c = 50\%$. It is seen that if 5 Mbps is the mean traffic requirement, then the cellular tier can cover up to a distance of 90 m whereas the D2D and femto can cover up to 30 and 21 m, respectively, which means that beyond a distance of 90 m, it is fair to switch over to either D2D or femto tier. The conclusion is that the cellular tier has a higher coverage distance whereas with less coverage but with higher throughput can be served by D2D and femto tiers. As seen that after a distance of 160 m, the cellular throughput becomes very small and thus our conclusion

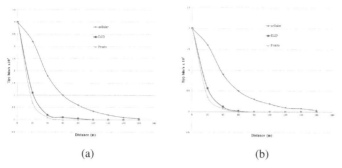

(a)   (b)

Figure 5.7   (a) Throughput under 50% cellular load. (b) Throughput under 75% cellular load.

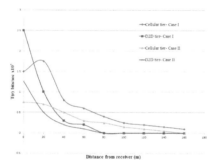

Figure 5.8   Throughput for a D2D-based cellular system.

is that we should opt for D2D and femto modes toward the cell edge whatever be the mean traffic requirement.

Figure 5.7(b) shows the results when $\beta_c = 75\%$. It is seen that the throughput of each mode has decreased due to the increase in β_c. Based on the average traffic requirement, 5 Mbps cellular coverage distance has fallen down from 90 to 60 m, whereas the D2D and femto tiers' coverage distance for 5 Mbps requirement has fallen down to 21 and 18 m, respectively. We conclude that the effect of higher loads has more detrimental effect on cellular tier whereas the impact on D2D and femto tier is less. Thus, it can be said that at higher loads, offloading to D2Ds and femto may be given preference.

Figure 5.8 shows the throughput obtained while using sub-band k, when only D2D is the part of HetNet and no femto involvement is there. It depicts the variation in the throughput as the distance of separation between the transmitter and the receiver within each tier varies under two cases i) $\bar{\beta}_c = 50\%$ and ii) $\bar{\beta}_c = 75\%$. In both cases, $v_d = 25\%$. It is observed that for the first case, D2Ds have higher throughput 25 Mbps at the proximity to 2.5 Mbps for distance 40 m, and then there is much degradation and it dies out at 80 m. It is seen that cellular user throughput on the other hand is 14 Mbps at the proximity, at 100 m, it reduces to 2.5 Mbps, and gradually goes on decreasing covering nearly 160 m. The observation is that the same 2.5 Mbps is reached at a separation distance of 40 m for D2D tier, whereas cellular tier users have throughputs higher than 2.5 Mbps even till a separation distance 100 m. For the second case, a similar trend is observed; however, both tiers' throughputs are of reduced value. It can be concluded that D2D tier fetches higher throughput with shorter coverage span, whereas cellular tier fetches larger coverage span with lesser througput. Further conclusion is that $\bar{\beta}_c$ has detrimental effect on both tiers.

6

Energy-efficient OFDMA Radio Access Networks

Energy-efficient network design is an important concern for cellular operators for environmental and economic reasons. Several approaches are being considered by the cellular operators to reduce energy consumption both at component and at a network level [CZB+10]. It is known that most of the power consumption in cellular networks is at the radio access network (RAN), mainly at the base stations (BSs) [HBB11, MBCM09]. Significant amount of energy-saving (ES) can be achieved by an efficient design of the network during planning and management phases [CZB+10, RFF09]. BSs are usually deployed considering peak traffic and future traffic growth. However, due to spatial and temporal variations in traffic conditions, most of the time, the BSs are largely underutilized [Hua10, FJL+13]. It is studied that about 30% of the time in a day, the traffic is below 10% of the peak [SKYK11]. Power consumption at the BS can be broadly divided into fixed and dynamic. Fixed is due to cooling, signal processing, etc., which constitutes 50% of the total power consumption [FBZ+10]. The dynamic part is mainly owed to the radio frequency (RF) transmission, which is a function of traffic load. Low traffic load situations cause severe degradation of area energy efficiency (AEE) [number of bits transmitted per Joule per unit area (unit: bits/Joule/m^2)]. Therefore, putting less loaded BSs in sleep mode (SLM) during low traffic hours is a promising way toward increasing AEE [CZB+10, FJL+13]. In SLM, the low loaded BSs are switched OFF and users' association is made to BSs which can lead to overall improvement of network performance. For the purpose of activating/deactivating the BSs, distributed and centralized ES functionalities have been included in Third Generation Partnership Project (3GPP) standard [3GP11].

Although SLM has the potential to improve AEE, it must be considered along with its side effects. The BS locations and RAN parameters are optimized in the initial stage to meet the system performance metrics such as coverage, capacity, overlap, quality of service (QoS), etc. [LL02, CGKW97, LLL98, MTR00]. Putting some BSs to SLM is naturally expected to affect these critical system performance indicators.

The subset of BSs to be put into SLM should be selected such that it does not affect coverage, capacity, and overlap requirements. But, these metrics are tightly coupled with each other due to intertwined interference effects. Further, it may be noted that these requirements of the network are a function of the network traffic variations [CZB+10, AIIE13].

6.1 Multi-objective Optimization in for Energy Savings

Multi-objective optimization is a natural requirement for planning radio network [LL02, CGKW97, LLL98, MTR00]. Identifying BS locations is a combinatorial optimization problem, and the solutions provided are meta-heuristic in nature [GSJZ09]. However, not much consideration was required to be paid toward energy efficiency objective as it is a rather newer requirement in the domain of cellular network planning. Further, dynamic network planning has evolved as a relatively newer requirement due to increasingly dynamic nature of the network traffic in both space and time domains. The dynamic reconfiguration of RAN requires re-initialization of the BS parameters, which is addressed in this part.

6.2 System Model

The details of the system model described are given in [CD16]. $\mathcal{B}$ is considered to be the set of $N_\mathcal{B}$ sectors located inside the considered geographical region $\mathbb{D} \subset \mathbb{R}^2$ with area $\mathcal{A}_\mathbb{D}$. The geographical region $\mathbb{D}$ is divided into n small rectangular grids each with area dA_i, where i grid index. The users are spatially distributed within the network with some distribution $p(.)$ such that $\int_\mathbb{D} p(i) \, dA_i = 1$. The network uses orthogonal frequency division multiple access (OFDMA) as the air interface with the frequency reuse of unity. The available total system bandwidth B Hz consists of N_{sc} number of subchannels each with bandwidth Δf_{sc} Hz.

The signal power received at location i from sector j is modeled as

$$P_{r,ij} = P_{tj} \, G_A(\theta_{ij}, \phi_{ij}) \, PL(d_{ij}, \alpha, f_c) \, |h_{ij}|^2 \, \chi_{ij}, \tag{6.1}$$

where P_{tj} is transmit power of the j-th sector, antenna gain [m2108]

$$G_A(\theta_{ij}, \phi_{ij}) = -\min[-(A_{\theta_{ij}} + A_{\phi_{ij}}), A_m], \tag{6.2}$$

where,

$$A_{\theta_{ij}} = -\min\left[\left(\frac{\theta_{ij}}{\theta_{3dBj}}\right)^2, A_m\right], A_{\phi_{ij}} = -\min\left[\left(\frac{\phi_{ij} - \phi_{\text{tilt}j}}{\phi_{3dBj}}\right)^2, A_m\right] \quad (6.3)$$

are the horizontal and vertical antenna patterns, respectively [m2108], θ_{ij} and ϕ_{ij} are the azimuth and elevation angle between the j-th sector antenna and location i, respectively, $\phi_{\text{tilt}j}$ is the vertical tilt angle of the j-th sector antenna, θ_{3dBj} and ϕ_{3dBj} are the 3-dB beamwidth of horizontal and vertical antenna patterns of the j-th sector antenna and A_m is the maximum attenuation [m2108]. The elevation angle is $\phi_{ij} = \arctan\left(\frac{H_{tj} - H_{ti}}{d_{ij}}\right)$, where H_{tj} and H_{ti} are the heights of the j-th sector and user at location i, respectively. Nakagami-m distributed random variable $|h_{ij}|$ represents the envelope of small-scale fading gain between location i and j-th sector, so the power of fast fading $|h_{ij}|^2$ follows Gamma distribution. Log-normally distributed random variable χ_{ij} represents the shadowing component between location i and j-th sector, i.e., $\chi_{ij} = \exp(\eta \xi_{ij})$, where ξ_{ij} is a Gaussian random variable (in dB) with zero mean and variance $\sigma_{\xi_{ij}}^2$ and $\eta = \frac{\ln(10)}{10}$. Path loss component $PL(d_{ij}, \alpha, f_c)$ is denoted as a function of the distance between the j-th sector and location i (d_{ij}), pathloss exponent α, and carrier frequency f_c.

6.2.1 SINR

The received signal-to-interference plus noise ratio (SINR), which is a function of shadowing (χ), small-scale fading (h), and activity status of interfering cells (v), for a user at location i associated with sector j is

$$\gamma_{k,ij}(\chi, h, v)$$
$$= \frac{P_{tkj} G_{ij}(\theta_{ij}, \phi_{ij}) PL(d_{ij}, \alpha, f_c) \chi_{ij} |h_{kij}|^2}{\sum_{g \neq j, g \in \mathcal{B}_{\text{on}}} v_{kg} P_{tkg} G_{ig}(\theta_{ig}, \phi_{ig}) PL(d_{ig}, \alpha, f_c) \chi_{ig} |h_{kig}|^2 + \Delta f_{sc} N_0},$$
$$(6.4)$$

where $\mathcal{B}_{\text{on}} \subseteq \mathcal{B}$ is set of active sectors, P_{tkj} and P_{tkg} denote the transmit power of desired sector j and interfering sector g, respectively, on subchannel k, and N_0 is the noise power spectral density in W/Hz. The j-th interfering sector's activity status on subchannel k is modeled as a binary random variable v_{kj} such that $Pr[v_{kj} = 1] = \beta_j$ and $Pr[v_{kj} = 0] = 1 - \beta_j$, where β_j denotes the activity factor of the j-th sector and Pr indicates the probability. The factor β_j can be seen as the fraction of resources occupied in cell j. The resource occupancy in an individual sector is dependent upon the offered traffic demand and the SINR statistics in that cell. Calculation of β_j is discussed in Section 6.3.

6.2.2 Traffic

Streaming traffic is considered where the blocking probability is taken as the QoS metric. The inter-arrival time of the calls per unit area $\lambda(i)$ is independently distributed exponential random variable with mean $1/\lambda$, while the duration of the call with mean $1/\mu(i)$ at location i is represented by $\mu(i)$, which is independently distributed exponential random variable. The traffic demand density of the call at location i is $\rho(i) = \lambda(i)/\mu(i)$ (Erlang/m^2).

6.3 Network Coverage, Overlap, ASE, and APC

The objective functions: network coverage, overlap probability, area spectral efficiency (ASE), and area power consumption (APC) are presented. The interference distribution is obtained as a function of active set of sectors $\mathcal{B}_{\text{on}}$, transmit powers $\mathcal{P} = [P_{t1}, P_{t2}, ..., P_{tN_B}]$, sector antenna's vertical tilt angles $\phi = [\phi_{\text{tilt1}}, \phi_{\text{tilt2}}, ..., \phi_{\text{tilt}N_B}]$, sector antenna heights $\mathcal{H} = [H_{t1}, H_{t2}, ..., H_{tN_B}]$, and resource occupancy $\beta = [\beta_1, \beta_2, ..., \beta_{N_B}]$ as the mean $\mu_{P_{I_i}} = \mu_X$ and variance $\sigma_{P_{I_i}}^2 = \sigma_X^2$ of the total interference power are obtained as a function of these parameters. For notational simplicity, these parameters are not used below.

6.3.1 Network Coverage

The probability that the user at location i is under the coverage of sector j is obtained, as follows:

$$Pr(\gamma_{ij} \geq \Gamma_{\min}, P_{r,ij} \geq P_{r,\min})$$

$$= Pr\left(\frac{P_{r,ij}}{\Gamma_{\min}} \geq P_{I_i}, P_{r,ij} \geq P_{r,\min}\right)$$

$$= Q\left(\frac{\ln(P_r) - \mu_{P_{r,ij}}}{\sigma_{P_{r,ij}}}\right)$$

$$+ \int_{P_{r,\min}}^{\infty} Q\left(\frac{\ln(P_r/\Gamma_{\min}) - \mu_{P_{I_i}}}{\sigma_{P_{I_i}}}\right) p_{P_{r,ij}}(P_r) dP_r, \qquad (6.5)$$

where $Q(x) = \frac{1}{\sqrt{2\pi}} \int_x^\infty e^{-\frac{t^2}{2}} dt$. By using Chernoff bound for Q-function, (6.5) can be approximated, as follows:

$$\hat{P}r(\gamma_{ij} \geq \Gamma_{min}, P_{r,ij} \geq P_{r,min}) = e^{\frac{\left(\frac{\ln(P_{r,min}) - \mu_{P_{r,ij}}}{\sigma_{P_{r,ij}}}\right)^2}{2}}$$

$$+ \frac{\sigma_{P_{I_i}} e^{\frac{(\mu_{P_{I_i}} - \mu_{P_{r,ij}})^2}{2(\sigma_{P_{I_i}}^2 - \sigma_{P_{r,ij}}^2)} + \left(\frac{\ln(\frac{P_{r,min}}{\Gamma_{min}}) - \mu_c}{8\sigma_c^2}\right)^2}}{\sqrt{\sigma_{P_{I_i}}^2 - \sigma_{P_{r,ij}}^2}}.$$

The fraction of total area covered by the sector j is obtained by averaging over the locations, i.e.,

$$P_{Cov,j}(\Gamma_{min}, P_{r,min}) = \sum_{i \in \mathbb{D}} \hat{P}r(\gamma_{ij} \geq \Gamma_{min}, P_{r,ij} \geq P_{r,min}) p(i) \, dA_i. \quad (6.6)$$

The probability that the location i is covered by at least one sector can be written as,

$$P_{Cov,i}(\Gamma_{min}, P_{r,min}) = 1 - \prod_{j \in \mathcal{B}_{on}} \left[1 - \hat{P}r(\gamma_{ij} \geq \Gamma_{min} P_{r,ij} \geq P_{r,min}) \right]. \quad (6.7)$$

Finally, the coverage probability for the entire region can be obtained by

$$f_{COV}(\mathcal{B}_{on}, \boldsymbol{\beta}, \boldsymbol{P}, \boldsymbol{\phi}, \mathcal{H}) = \sum_{i \in \mathbb{D}} P_{Cov,i}(\Gamma_{min}, P_{r,min}) \, p(i) \, dA_i. \quad (6.8)$$

6.3.2 Overlap Probability

A location i is said to be under the coverage of b sectors if the received SINR from any of b of active sector set is greater than the minimum SINR threshold Γ_{min} and is less than that for remaining sectors.

The overlap probability can be represented in terms of coverage and outage probabilities. The probability that the location i is covered by b sectors is obtained as given in (6.10).

In (6.10), Q_{x_1} represents the outage probability with respect to the x_1-th sector, i.e., $Q_{x_1} = Q\left[\frac{\mu_{\gamma_{ix_1}} - \ln(\Gamma_{min})}{\sigma_{ix_1}}\right]$. Similarly, the probability that the location is covered by at least b number of sectors can be obtained as

$P_{OL,i}^{(>b)} = \sum_{b'=b}^{N} P_{OL,i}^{(b')}$. The fraction of total area covered by b sectors is obtained by

$$f_{\mathrm{OL}}(\mathcal{B}_{\mathrm{on}}, \boldsymbol{\beta}, \boldsymbol{\mathcal{P}}, \boldsymbol{\phi}, \boldsymbol{\mathcal{H}}) = \sum_{i \in \mathbb{D}} P_{OL,i}^{(b)} \, p(i) \, dA_i. \tag{6.9}$$

$$P_{OL,i}^{(b)} = \sum_{x_1=1}^{N-b+1} (1 - Q_{x_1})$$

$$\left[\sum_{x_2=x_1+1}^{N-b+2} (1 - Q_{x_2}) \left[\cdots \cdots \left[\sum_{x_{N-2}=x_{N-3}+1}^{N-1} (1 - Q_{x_{N-2}}) \right. \right. \right.$$

$$\left. \left. \left. \left[\sum_{x_{N-1}=x_{N-2}+1}^{N} (1 - Q_{x_{N-1}}) \sum_{x_{N+1} \neq x_1, x_2, x_3, \ldots, x_N}^{N} Q_{x_{N+1}} \right] \right] \cdots \cdots \right] \right]. \tag{6.10}$$

For example, when $N_{\mathcal{B}} = 4$ and $b = 2$,

$$P_{OL,i}^{(2)} = (1 - Q_1) \left[\sum_{x_2=2}^{2} (1 - Q_{x_2}) \left[\sum_{x_3=x_2+1}^{3} (1 - Q_{x_3}) \right. \right.$$

$$\left. \left. \left[\sum_{x_4=x_3+1}^{4} (1 - Q_{x_4}) \sum_{x_5 \neq x_1, x_2, x_3, x_4}^{4} Q_{x_5} \right] \right] \right].$$

$$= (1 - Q_1)(1 - Q_2)Q_3 Q_4 + (1 - Q_1)(1 - Q_3)Q_2 Q_4$$
$$+ (1 - Q_1)(1 - Q_4)Q_2 Q_3 + (1 - Q_2)(1 - Q_3)Q_1 Q_4$$
$$+ (1 - Q_2)(1 - Q_4)Q_1 Q_3 + (1 - Q_3)(1 - Q_4)Q_1 Q_2.$$

6.3.3 Area Spectral Efficiency

If N_L is the number of modulation and coding scheme (MCS) levels and the corresponding SINR thresholds are $\Gamma_1, \Gamma_2, \ldots, \Gamma_{N_L-1}, \Gamma_{N_L}$, in the long term, the fraction of users belonging to a particular range of SINR thresholds can be written as

$$w_j(l) = Pr(\gamma_{ij} \geq \Gamma_l, P_{r,ij} \geq P_{r,\min}) - Pr(\gamma_{ij} \geq \Gamma_{l+1}, P_{r,ij} \geq P_{r,\min}).$$

If b_l is the number of bits transmitted per Hz by using l-th MCS, then the average spectral efficiency can be written as $\overline{T}_j = \sum_{l=1}^{N_L} w_j(l) \cdot b_l$.

The average throughput (bits/s) achieved by the j-th cell is computed as $C_j = B.\overline{T}_j = B.\sum_{l=1}^{N_L} w_j(l). b_l$. The ASE (b/s/Hz/m^2) is obtained by

$$f_{\text{ASE}}(\mathcal{B}_{\text{on}}, \boldsymbol{\beta}, \boldsymbol{\mathcal{P}}, \boldsymbol{\phi}, \boldsymbol{\mathcal{H}}) = \frac{1}{BA_{\mathbb{D}}} \sum_{j \in \mathcal{B}_{\text{on}}} C_j. \tag{6.11}$$

6.3.4 Blocking Probability

The average blocking probability in the j-th cell is obtained as the weighted sum of all class-wise blocking probabilities, i.e., $P_{bj} = \frac{1}{\sum_{c=0}^{N_{sc}} g_j(c)} \sum_{l=1}^{N_L} w_j(l) \sum_{c=N_{sc}-n_{sc}(l)+1}^{N_{sc}} g_j(c)$. The fraction of bandwidth utilized in cell j can be computed as $\beta_j = \frac{1}{N_{sc}} \sum_{c=1}^{N_{sc}} \frac{1}{c} \sum_{l=1}^{N_L} \rho_j w_j(l).n_{sc}(l).g_j(c-n_{sc}(l))$.

The feasible load vector that satisfies the blocking probability requirements, (i.e., $P_{bj} \le P_{b,\max}$) can be written as

$$\mathcal{F}\big(\boldsymbol{\beta}(\mathcal{B}_{\text{on}}, \boldsymbol{\beta}, \boldsymbol{\mathcal{P}}, \boldsymbol{\phi}, \boldsymbol{\mathcal{H}})\big)$$

$$= \left\{ \boldsymbol{\beta} \middle| \beta_j = \frac{1}{N_{sc}} \sum_{c=1}^{N_{sc}} \frac{1}{c} \sum_{l=1}^{N_L} \rho_j(l).n_{sc}(l).g_j(c - n_{sc}(l)), \right.$$

$$\left. 0 \le \beta_j \le 1 - \epsilon, \ P_{bj} \le P_{b,\max}, \ \forall j \in \mathcal{B}_{\text{on}} \right\}, \tag{6.12}$$

where $\epsilon > 0$ is an arbitrarily small positive number. The load vector is solution of the system

$$\boldsymbol{\beta}^*(\mathcal{B}_{\text{on}}, \boldsymbol{\beta}, \boldsymbol{\mathcal{P}}, \boldsymbol{\phi}, \boldsymbol{\mathcal{H}}) = \mathcal{F}\big(\boldsymbol{\beta}(\mathcal{B}_{\text{on}}, \boldsymbol{\beta}, \boldsymbol{\mathcal{P}}, \boldsymbol{\phi}, \boldsymbol{\mathcal{H}})\big). \tag{6.13}$$

Since the load vector, $\boldsymbol{\beta}$ is a complex function of load in other cells, the solution of (6.13) has a fixed point in $[0, 1)^{|\mathcal{B}_{\text{on}}|}$. The solution can be iteratively obtained using the fixed point iteration method [SY12].

6.3.5 Area Power Consumption

The power consumption at sector j is modeled as [AGG$^+$11]

$$P_{Cj} = \frac{N_{\text{TRX}}.(\beta_j.\frac{P_{tmax,j}}{\eta_{\text{eff}}} + P_{RF} + P_{BB})}{(1 - \sigma_{DC})(1 - \sigma_{MS})(1 - \sigma_{cool})}, \tag{6.14}$$

where N_{TRX} denotes the number of TRX chains of the sector j, $P_{tmax,j}$ denotes the maximum RF output power of sector j at peak load, and η_{eff}

denotes the power amplifier (PA) efficiency. σ_{DC}, σ_{MS}, and σ_{cool} are the loss factors due to DC–DC power supply, main supply, and cooling, respectively. The APC (W/m^2) by all the active BSs ($\mathcal{B}_{\text{on}}$) is obtained by

$$f_{\text{APC}}(\mathcal{B}_{\text{on}}, \beta, \mathcal{P}, \phi, \mathcal{H}) = \frac{1}{\mathcal{A}_{\mathbb{D}}} \sum_{j \in \mathcal{B}_{\text{on}}} P_{Cj}. \qquad (6.15)$$

The objective functions: network coverage (6.8), overlap (6.9), ASE (6.11), and APC (6.15), and the blocking probability are coupled with each other through co-channel interference (CCI).

6.4 Multi-objective Optimization Framework

Given the traffic demand density ρ (Erlang/m^2), the objective is to find the set of active sectors ($\mathcal{B}_{\text{on}}$), and sector antenna's vertical tilt angles (ϕ), sector transmit powers ($\mathcal{P}$), and sector antenna heights ($\mathcal{H}$) that jointly maximizes the network coverage (f_{COV}), ASE (f_{ASE}), and minimizes APC (f_{APC}) and the overlap (f_{OL}) while satisfying target QoS (blocking probability) requirements. Let $\mathbf{x} = [x_1, x_2, x_3, x_4, x_5]^T = [\mathcal{B}_{\text{on}}, \beta, \phi, \mathcal{P}, \mathcal{H}]^T$ and $g_1(\mathbf{x}) = -f_{\text{APC}}(\mathcal{B}_{\text{on}}, \beta, \mathcal{P}, \phi, \mathcal{H})$, $g_2(\mathbf{x}) = f_{\text{ASE}}(\mathcal{B}_{\text{on}}, \beta, \mathcal{P}, \phi, \mathcal{H})$, $g_3(\mathbf{x}) = f_{\text{COV}}(\mathcal{B}_{\text{on}}, \beta, \mathcal{P}, \phi, \mathcal{H})$, and $g_4(\mathbf{x}) = -f_{\text{OL}}(\mathcal{B}_{\text{on}}, \beta, \mathcal{P}, \phi, \mathcal{H})$. The multi-objective optimization problem is formulated as

$$\max_{\mathbf{x}} \ \mathbf{g}(\mathbf{x}) = [g_1(\mathbf{x}), g_2(\mathbf{x}), g_3(\mathbf{x}), g_4(\mathbf{x})]^T \qquad (6.16)$$

$$\text{Subject to: } C1 : P_{bj} \leq P_{b,\max} \ j \in \mathcal{B}_{\text{on}};$$
$$C2 : g_3(\mathbf{x}) \geq f_{\text{COV}\,\min}$$
$$C3 : \mathcal{P}_{\min} \leq x_3 \leq \mathcal{P}_{\max}$$
$$C4 : \phi_{\min} \leq x_4 \leq \phi_{\max}$$
$$C5 : \mathcal{H}_{\min} \leq x_5 \leq \mathcal{H}_{\max}.$$

Here, constraint $C1$ represents the blocking probability requirements. Constraint $C2$ represents the minimum coverage requirement. Constraints $C3$, $C4$, and $C5$ represent the range of transmitting power, tilt, and height, respectively, at the individual sector.

Finding the optimal set of sector and RAN parameters among a large number of combinations is a complex combinatorial problem. The four objectives in (6.16) are conflicting as they are coupled with each other due to the complex relationship among the variables. As there is no single solution that maximizes all the objectives simultaneously, in general, there is no global optimum to the multi-objective optimization problem in (6.16). Let $\mathcal{G} = \{g(\mathbf{x}), \forall \mathbf{x}\}$ be the objective set which contains all possible combinations of objective values. There will be multiple solutions called *Pareto optimal set*. A solution is said to be Pareto optimal, if none of the objective functions can be improved in value without degrading some of the other objective values [AJG05].

6.4.1 Genetic Algorithm-based Multi-objective Optimization

Evolutionary algorithms such as genetic algorithm (GA) can find the solutions to the problem described above in a relatively short time because their time complexity is not directly related to the dimension of the problem. Another important feature of GA is that it is well suited for dynamic environments as well. There are two different types of approaches that can be used for dynamically adapting GA: *search-based* and *memory-based* approaches [BAB05, MK00]. In memory-based approach, the set of solutions from the past adaptation is used as a candidate solution for the next adaptation. For instance, if the traffic load at a particular time of the day is the same as that of previous day, then the solutions of the previous day can be used as a candidate solution set. But, if the traffic condition is unpredictable and different from the previous day traffic, search-based approaches such as *random immigrants* [JB05] can be used. In random immigrants method, for each generation, the worst individuals are replaced by the randomly generated individuals in order to increase the diversity in the population. These approaches need only small adjustments in reproduction operator in GA. The dynamics of GA can be implemented both in *online* mode and *offline* mode. Another important feature of GA is that the speed can be improved by parallel implementation of GA [Deb01]. The centralized self-organizing network (SON) functionalities such as load balancing, hand-over parameter optimization, interference control, capacity and coverage optimization, etc., are being considered in Long Term Evolution (LTE) networks [3GP08]. Further, centralized ES functionalities have been added in 3GPP standard [3GP11]. Therefore, the sectors and the corresponding RAN parameters can be identified for SLM by exploiting the features of centralized ES and SON functionalities.

The approach described here is called *a posteriori method*, where first the set of Pareto optimal solutions is obtained and then the final solution is selected based on the preferences. Each solution in the Pareto optimal set has certain tradeoffs between the objectives. As there will be multiple solutions, from the Pareto solution set, the network operator has an opportunity to select appropriate BS and RAN parameter configuration according to their needs. In this work, we obtain the Pareto optimal solutions using non-dominated sorting genetic algorithm (NSGA).

In order to reduce the complexity, the above optimization problem is solved in two steps. In the first step, given the traffic demand density ρ, the set of active sectors is found $(\mathcal{B}_{\text{on}})$ with fixed RAN parameter configuration. In the second step, we obtain the sector transmit powers $(\mathcal{P})$, sector antenna heights $(\mathcal{H})$ and antenna tilt angles (ϕ) for the set of active sectors which are obtained from the first step.

6.4.2 Set of Active Sectors

Given the traffic demand density ρ, with fixed RAN parameter configuration $\big($ antenna tilt angles $\boldsymbol{\Phi} = [\Phi_{\text{tilt}1}, \Phi_{\text{tilt}2}, ..., \Phi_{\text{tilt}N_{\mathcal{B}}}]$, sector antenna heights $\mathbf{H} = [H_{t1}, H_{t2}, ..., H_{tN_{\mathcal{B}}}]$, and sector transmit powers $\mathbf{P} = [P_{t1}, P_{t2}, ..., P_{tN_{\mathcal{B}}}]\big)$, the set of sectors that maximize coverage and ASE, and minimize the APC and overlap is

$$\max_{\mathbf{x}(1)} \ \mathbf{g}(\mathbf{x}) = [g_1(\mathbf{x}), g_2(\mathbf{x}), g_3(\mathbf{x}), g_4(\mathbf{x})]^T, \text{Subject to:- } C1, C2.$$

The optimal set of active sectors is obtained by searching the entire search space

$$\mathcal{T}_{\mathcal{B}} = \{\mathcal{B}_{\text{on}}^1, \mathcal{B}_{\text{on}}^2, ..., \mathcal{B}_{\text{on}}^{(N_{\text{SP}}^{\mathcal{T}_{\mathcal{B}}})}\} \tag{6.17}$$

which contains all possible active sector configurations with size $N_{\text{SP}}^{\mathcal{T}_{\mathcal{B}}} = |\mathcal{T}_{\mathcal{B}}| = 2^{N_{\mathcal{B}}} - 1$.

6.4.3 RAN Parameters

For a set of active sectors $(\mathbf{B}_{\text{on}})$, the set of RAN parameters that maximize coverage and ASE, and minimize APC and overlap is

$$\max_{\mathbf{x}} \ \mathbf{g}(\mathbf{x}) = [g_1(\mathbf{x}), g_2(\mathbf{x}), g_3(\mathbf{x}), g_4(\mathbf{x})]^T$$

$$\text{Subject to: } C1, C2, C3, C4, C5.$$

Let $\phi_{\text{tilt}j} = \{\phi_{\text{tilt}j_{\min}}, ..., \phi_{\text{tilt}j_{\max}}\}$, $P_{tj} = \{P_{tj_{\min}}, ..., P_{tj_{\max}}\}$, and $H_{tj} = \{H_{tj_{\min}}, ..., H_{tj_{\max}}\}$ be the set of values of tilt angles, transmit power, and heights, respectively, available at sector j. Let $K_{P_t}^j = |P_{tj}|$, $K_{\phi_{\text{tilt}}}^j = |\phi_{\text{tilt}j}|$, and $K_{H_t}^j = |H_{tj}|$ be the cardinality of $\phi_{\text{tilt}j}$, P_{tj}, and H_{tj}, respectively. A particular combination of set of RAN parameters is denoted as

$$\mathcal{R} = \{X, Y, Z | X \in \phi_{\text{tilt}j}, Y \in P_{tj}, Z \in H_{tj}\}, \mathcal{R} \in \mathcal{T}_\mathcal{R}. \tag{6.18}$$

The optimal RAN parameter configuration is obtained by searching the entire search space

$$\mathcal{T}_\mathcal{R} = \{\mathcal{R}^1, \mathcal{R}^2, ..., \mathcal{R}^{(N_{\text{SP}}^{\mathcal{T}_\mathcal{R}})}\}. \tag{6.19}$$

6.5 Results

The dense Urban Micro-cell (UMi) network scenario with an inter-site-distance of 200 m, is used for evaluating the proposed framework. Wrap around model is used to incorporate equal CCI at the edge regions. The user locations are assumed to be uniformly distributed within the network, i.e., $p(i) = \frac{1}{\mathcal{A}_\mathbb{D}}$, $\forall i$. The total system bandwidth is of 10 MHz composed of 600 useful subchannels, each with bandwidth 15 kHz is considered. Streaming call requests with a rate requirement (R_{req}) of 128 kbps are taken. The maximum allowable blocking probability in a cell is set to $P_{b,\max} = 0.02$. Further, homogeneous traffic distribution is assumed (i.e., $\rho(i) = \rho$, $\forall i$). The power consumption parameters are shown in Table 6.1 are as in [AGG+11]. The

Table 6.1 System parameters used for evaluation of multi-objective optimization framework

Parameter	Value
Carrier frequency	2.5 GHz
Bandwidth (B)	10 MHz
Subcarrier spacing (Δf_{sc})	15 KHz
Number of MCS levels (N_L)	11
3-dB vertical beamwidth (ϕ_{3dB})	15 Deg.
3-dB horizontal beamwidth (θ_{3dB})	70 Deg.
Antenna gain (boresight)	17 dBi
Sector antenna height (H_t)	20 m
Vertical tilt angle (ϕ_{tilt})	12 Deg.
Thermal noise level (N_0)	-104 dBm/10 MHz
Minimum required SINR threshold ($\Gamma_{\min}$)	-10 dB
Target blocking probability ($P_{b,\max}$)	2%
Std. of shadow fading (σ)	4 dB

Table 6.2   Power consumption parameters used for evaluation of multi-objective optimization framework

Parameter	Value
N_{TRX}	1
PA power consumption:	
P_{max}	20.0 W (43 dBmW)
Back-off	8 dB
PA Efficiency η_{eff}	31.1 %
Total PA $P_{PA} = \frac{P_{\mathrm{max}}}{\eta_{\mathrm{eff}}}$	**64 W**
RF power consumption:	
P_{TX}, P_{RX}	6.8 W, 6.1 W
Total RF $P_{RF} = P_{TX} + P_{RX}$	**13 W**
Baseband power consumption:	
P_{BB}	**29.5 W**
Loss factors: $\sigma_{DC}, \sigma_{MS}, \sigma_{cool}$	7.5%, 9.0%, 10.0%
Total P_{in}	**140.5 W**

number of transmission/reception (TRX) chains is assumed to be one. The non-line of sight (NLoS) pathloss model for UMi scenario is as per [m2108]. The RAN parameters are used following [m2108]. The horizontal and vertical antenna pattern parameters are considered as in [m2108]. Minimum required received power ($P_{r,\mathrm{min}}$) and minimum required SINR (Γ_{min}) thresholds are assumed to be -102 dBmW (10 MHz) and -10 dB, respectively. The SINR thresholds for 15 different MCS levels for single-input single-output (SISO) Rayleigh fading scenario are used with the target block error rate (BLER) of 10^{-1} [MWI+09].

Figure 6.1 shows the effect of selecting solutions with the goal of achieving an individual objective on cost functions and ES. Figure 6.1(a) shows the APC when the solution is selected with for an individual objective function against varying traffic demand density. The corresponding ASE, % overlap, % coverage, and ES are shown in Figures 6.1(b–d) and 6.2, respectively.

It can be noted from Figure 6.2 that the solution with the objective of minimizing APC maximizes ES. However, it is achieved at the cost of significant reduction in ASE, and coverage. The % overlap increases when the solution is selected with the objective of minimizing APC instead of the case

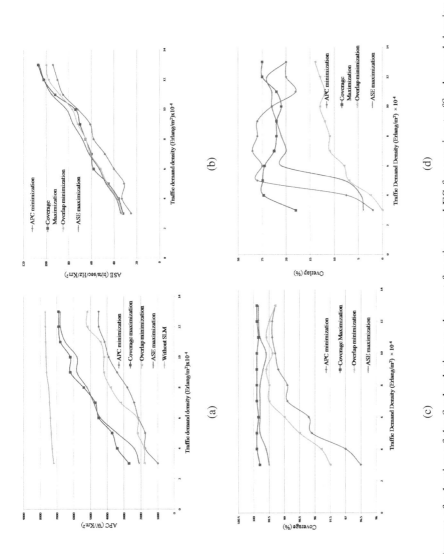

Figure 6.1 Impact of selection of the final solution on the cost functions and ES for varying traffic demand density. (a) APC, (b) ASE, (c) Coverage and (d) Overlap.

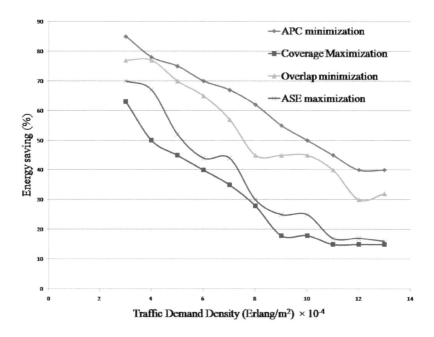

Figure 6.2   Impact of selection of the final solution on ES.

where the solution is selected with the objective of maximizing ASE. It can also be observed that when overlap minimization is chosen, then the possible ES is close to the case where the solution is selected with the objective of minimizing APC. The ASE is also nearly equal to the case where the solution is selected with the objective of maximizing ASE.

7

QoS Provisioning in OFDMA Networks

Wireless Internet traffic is expected to grow tremendously [Cis17] at a compound annual growth rate of 47% between 2016 and 2021 [Cis17]. In order to accommodate this exponential growth of data traffic, broadband wireless access (BWA) networks, such as long-term evolution (LTE), Worldwide Interoperability for Microwave Access (WiMAX), Fifth Generation (5G), etc., have adopted a packet network architecture [STB09]. The last mile wireless access poses a serious challenge in providing the increasing demand for high data rates because of the spatio-temporal fluctuations in the wireless link between the base station (BS) and user equipment (UE). To compensate for these fluctuations, cross-layer methods such as link adaptation (LA), packet scheduling and radio resource allocation (PS-RRA), etc., have evolved which maximize the supported data rates by exploiting the delay tolerant capability of packet data traffic [AM13, STB09, HT09].

Although packet-based wireless networks can provide high throughput by taking advantage of multi-user diversity and the delay tolerance of data traffic, they are also expected to carry real-time (RT) voice and video as a significant fraction of the total traffic. Both voice and video are highly delay sensitive [Cis17]. The capacity of networks which carry these RT traffic is measured in terms of the number of active user connections with satisfied quality of service (QoS) guarantees [Rap01], which is in contrast to packet centric network, where the unit of measure is data rate (bits/second).

Radio resource allocation (RRA) algorithms for transmitting QoS-guaranteed traffic over packet-based radio access networks (RANs) [YMPM08, PKH$^+$08, P.12, PD15], aim toward efficient utilization of radio resources so as to maximize the number of actively connected users. Based on the user's signal-to-noise ratio (SNR), a variable number of radio resources are allocated in order to provide the user's required QoS [STB09, PD15]. Such methods are significantly different from earlier global system for mobile communication like technologies, where a fixed amount of radio resource was

207

granted for RT traffic. Since QoS guarantees require timely and successful delivery of a required number of packets, there is only a limited number of connections whose QoS can be satisfied for a given network condition. When excess traffic enters such networks, there ensues a catastrophic fall in the QoS [P.12]. QoS guarantees thus require a controller for admission of calls [Ahm05,Lopez2014].

To maximize RT traffic capacity with QoS guarantees, the call admission control (CAC) unit needs to admit the appropriate number of connections such that no more connections can be accommodated wherein all users' QoS is satisfied. Here QoS is defined by packet loss rate, average packet delay, etc. [IR08]. The maximum number of admissible users is dictated by the traffic handling capacity (efficiency) of RRA, and users' SNR, QoS demands, and also user distribution in the geographic area. As traffic demand and users' SNRs are dynamic, the maximum number of users admissible is not a fixed quantity and hence a detailed inspection of the interrelationship of the different components is required, which is provided in this chapter.

Many modern BWA networks, such as LTE, long term evolution-advanced (LTE-A), LTE-A pro, etc., use orthogonal frequency division multiple access (OFDMA) as the channelization scheme. In OFDMA networks, the bandwidth is partitioned into granulated physical resource blocks (PRBs), which are shared by the users [STB09]. Adaptive modulation and coding (AMC) is used to determine the data rate, which is as a function of users' SNR, supported by each PRB. This, in turn, can be used to derive the number of PRBs required by a user to support a given bit rate [3GP08]. Therefore, it is easily established that PRB requirement of different users may be different depending on their SNR and QoS demands. Furthermore, RT traffic does not begin transmission till the required number of PRBs becomes available [STB09, P.12]. As a result, some PRBs may remain free even when the buffers are non-empty. Hence, the frequency domain packet scheduling/RRA algorithms of OFDMA networks can be mapped into multi-resource (due to multiple PRBs requirements) non-work conserving schedulers [GSTH08].

Accurate prediction of the QoS metrics supported by RRA algorithms in these networks has been studied in earlier works through extensive time-consuming offline simulations [P.12,PD15,YMPM08]. These simulations are cumbersome and cannot be used for RT QoS prediction as and when the network configuration changes. However, these estimates are required by CAC. In contrast, analytical models of RRA algorithms for a given set of

Table 7.1   List of symbols

N_{PRB}	Total number of PRBs available
N	Total number of users in the system
$\vec{\xi}$	Scheduler control parameter vector
ζ_k	Traffic density of class-k
λ_{p_k}	Average packet arrival rate of class-k
λ_c	Average call arrival rate
μ_c	Call holding time
τ	Duration of a TTI
$\mathbb{U}_k$	Tagged user of class-k
N_k	Number of users in class-k
$\gamma^u/\overline{\gamma^u}$	Instantaneous/average SNR of any user "u"
$\gamma_k/\overline{\gamma_k}$	Instantaneous/average SNR of class-k
$\gamma_k^u/\overline{\gamma_k^u}$	Instantaneous/average SNR of user "u" of class-k
$\mathcal{A}_{k,n}$	Number of class-k packet arrivals in TTI n
$\mathcal{B}_k$	Buffer of class-k
$b_{k,n-1}$	Length of buffer $\mathcal{B}_k$ at the beginning of TTI n
$C_{k,n}$	Server state or AMC mode of $\mathbb{U}_k$ in TTI n
sc_n	Prioritized class in TTI n
$N_{PRB,k}^n$	Number of PRB available to class-k in TTI "n"
R_ω	Number of PRB required with AMC mode ω
δ_k	Average packet delay of class-k
P_{D_k}	Packet drop probability of class-k
$\mathbb{T}_k/\mathbb{T}^u$	Throughput of class-k/user "u"
γ_k^{th}	Average SNR thresholds for user classification
γ_ω^b	Boundary of SNR interval for AMC mode ω

network parameters can predict QoS metrics with a higher timeliness and better accuracy. They can be thus used online, which makes them useful for efficient and dynamic CAC [Lopez2014,BD14,LHA06].

An analytical framework of a dynamic priority (DP) RRA algorithm is described in this chapter. The DP algorithm dynamically assigns the priority of resource assignment to multiple user classes (SNR-based) using a scheduler control parameter. The problem is mapped into a multi-class multi-server finite buffer queueing system, where the class-wise service processes, modeled using AMC, are non-deterministic. The scheduler behavior, buffer states, and AMC modes are combined in the network for which an augmented finite state Markov chain (FSMC) is designed. It is subsequently solved to derive QoS metrics such as average packet delay, probability of packet drop due to buffer overflow, average throughput, etc.

7.1 System Model of the Dynamic Priority Scheduler

The DP scheduler has been developed for the downlink of an OFDMA cellular network in which the system bandwidth of B Hz is divided into $\mathrm{N_{PRB}}$ orthogonal PRBs. Each PRB consists of $\mathrm{N_{sc}^P}$ subcarriers and spans a duration of τ ms, which is the scheduling interval or transmission time interval (TTI). All PRBs are orthogonally shared by N users which are uniformly deployed in a cell of area $\mathcal{C_R} \subset \mathbb{R}^2$. Users in $\mathcal{C_R}$ are served by a BS x_0 located centrally within $\mathcal{C_R}$. The signal received by user "u" $\in \mathcal{C_R}$ from its desired BS x_0 is $P_u^r = P_{x_0}^T G_u$, where $P_{x_0}^T$ is the transmit power of BS x_0 and G_u is the channel gain. G_u is the product of the distance-dependent path loss, L_u, and the power of small scale fading component, z_u, such that $G_u = L_u z_u$. Small scale fading and hence the envelope of the received signal is modeled to follow a generalized *Nakagami-m* distribution [Stu01]. Consequently, the power of small scale fading and therefore the instantaneous SNR γ^u of user "u" follows Gamma distribution.

The probability distribution function of γ^u is:

$$p_\gamma(\gamma^u) = \frac{m^m (\gamma^u)^{m-1}}{(\overline{\gamma^u})^m \Gamma(m)} \exp\left(-\frac{m\gamma^u}{\overline{\gamma^u}}\right), \qquad (7.1)$$

where $\overline{\gamma^u}$ is the average received SNR of user "u," $\Gamma(m) = \int_0^\infty x^{m-1} e^{-x} dx$ is the complete Gamma function, and "m," is the fading parameter of the Nakagami distribution.

User Classification

In this work, the average SNR $\overline{\gamma^u}$ is taken to be a function of the distance dependent path-loss only. So, if a user moves along a circle with the desired BS x_0 at the center, its received signal strength and received SNR remain unchanged. We can, therefore, draw concentric circles around the BS such that as the distance from the B increases, the radii of the circles increase, and the received SNR (power) decreases as shown in Figure 7.1 [BP03,BD15,CLCPRAHV14]. These concentric circles divide the area of the cell into K annular regions, as can be seen from Figure 7.1, wherein the average SNR of region k users falls in the range $\gamma_k^{th} \le \overline{\gamma^u} < \gamma_{k+1}^{th}$, [BP03,BPD16]. γ_k^{th} represents the average SNR thresholds for defining the SNR-based regions. We refer to these regions as classes. Although the framework does not consider interference from neighboring BSs, it can be extended to multi-cell scenarios with interference being treated as noise.

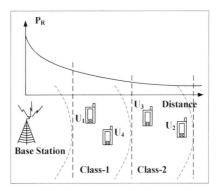

Figure 7.1 SNR-based classification of users.

We assume that packet arrivals from class-k follow a Poisson process with mean $\lambda_{p_k} = \zeta_k \lambda_p$. Here, λ_p is the total packet arrival rate in the area $\mathcal{C}_R$ and ζ_k is the percentage of packets or traffic density of class-"k," where $\sum_{k=1}^{K} \zeta_k = 1$. Thus, if there are N_k users in class-k, the packet arrival process for each user is Poisson with arrival rate $\left\lfloor \frac{\lambda_{p_k}}{N_k} \right\rfloor$. The assumption of Poisson packet arrivals, as in [YMPM08,IR08,PD15], and [TA14], allows us to exploit the simplicity of Markovian modeling leading to tractable mathematical solutions [TA14,LZG05,WHCW13,CLCPRAHV14]. Moreover, Poisson distribution has been widely used to model packet arrivals of voice over IP (VoIP) traffic [YMPM08,IR08,PD15] as well as of Internet traffic [NSVP11]. Furthermore, by considering Poisson arrivals, it becomes convenient to extend the model to Poisson networks as in [MSDP15] and the references therein.

In practical scenarios, packets of user "u" are stored in a dedicated buffer at the BS. This buffer is served according to the user's instantaneous channel conditions captured using AMC. A mathematical framework which involves the buffer states and AMC modes of individual users becomes unscalable with the increase in the number of users or the buffer lengths [RHHB09]. To overcome this problem, the analytical framework has been developed from the perspective of a unique user per class as in [PRB10] and [RHHB09]. The unique user of class-k, also called the "*tagged user*," is represented by $\mathbb{U}_k$. Users of class-k are assumed to have the same traffic properties, channel statistics, and QoS requirements as $\mathbb{U}_k$ [PRB10, RHHB09]. Hence, all packets of class-k users are stored in a buffer "$\mathcal{B}_k$" of length L_k at the BS and served according to the instantaneous AMC mode of $\mathbb{U}_k$ in that TTI.

Each AMC mode is a function of the underlying SNR of a user and corresponds to a modulation and coding scheme (MCS) which gives the number of bits that a user can transmit per symbol. Since the number of symbols that a PRB can carry is fixed, each AMC mode can be used to represent the number of PRBs that a user needs to transmit a packet of a given size [PD15,STB09,3GP08] in a TTI. We explain this using the following example.

In LTE, a PRB consists of 12 subcarriers and spans the duration of a TTI of 1 ms [STB09]. A TTI is divided into two 0.5 ms time slots, each of which accommodates 6/7 orthogonal frequency division multiplexing (OFDM) symbols depending on the length of the cyclic prefix [STB09]. So, each PRB consists of a maximum of $(2 \times 7 \times 12) = 168$ resource elements. In single-input single-output (SISO) systems, three OFDM symbols are reserved for control information and eight resource elements for pilot information [STB09]. So, the total number of resource elements available for data transmission is $(14-3) \times 12 - (8-2) = 126$. One resource element can carry one symbol [STB09]. So, the total number of symbols that a PRB can carry in a SISO system is 126, a fixed number. Now, let us consider that the chosen modulation scheme is quadrature phase shift keying (QPSK) and the coding rate is 1/2. So, the number of bits carried per symbol is $(2 \times \frac{1}{2} = 1)$. Therefore, the number of PRBs needed to transmit a packet of 320 bits is $\lceil \frac{320}{126} \rceil = 3$.

In this work, as the channel characteristic of all class-k users is the same as the class-k tagged user $\mathbb{U}_k$, their PRB requirement, in any TTI, is a function of the instantaneous SNR γ_k of $\mathbb{U}_k$. γ_k depends on $\mathbb{U}_k$'s average SNR $\overline{\gamma_k}$, which is given by:

$$\overline{\gamma_k} = \frac{1}{N_k} \sum_{u=1}^{N_k} \overline{\gamma_k^u}, \quad \forall u \in \text{class-}k, \qquad (7.2)$$

where $\overline{\gamma_k^u}$ is the average SNR of user "u" of class-k. The other major assumptions used in this work are:

A1: For each PRB, the channel response remains invariant within a TTI, but varies from one TTI to the next. This results in a block fading channel model that represents slowly varying channels [LZG05,WHCW13].

A2: The tagged user of a class sees the same frequency response across all PRBs. However, channel responses of the tagged user of different classes are different.

A3: Once a packet is assigned to the PRBs in any TTI, it has to be transmitted before the next TTI begins [PD15,STB09].

A4: Perfect channel state information is available at each receiver and is fed back using an instantaneous and error-free channel in the uplink [LZG05].

A5: Error detection [LZG05], [MZB01] at the receiver is perfect. Throughput calculation [LZG05] does not include packet overhead and cyclic redundancy check parity bits.

A6: Packets are sent without retransmissions and are dropped on incorrect reception [LZG05].

7.1.1 Modeling the Physical Layer (PHY) Layer

As a consequence of assumptions A1 and A2, the channel conditions of the tagged user $\mathbb{U}_k$ in all PRBs change at the same time to the same state. These channel fluctuations are captured using AMC, to model which the received SNR is partitioned into N_{AMC} intervals. Each SNR interval, for a given transmit power and target block error rate (BLER), supports a specific MCS and represents an AMC mode. At any time slot/TTI n, the BS chooses the AMC mode ω for user "u" if its instantaneous SNR $\gamma^u \in (\gamma^b_{\omega-1}, \gamma^b_\omega]$. Here, γ^b_ω and $\forall \omega$ are the boundaries of the SNR intervals that are used to determine the AMC modes. γ^b_ω are different from γ^{th}_k. To avoid information loss, no transmission is attempted when $\gamma^u < \gamma^b_1$. This condition gives rise to the AMC mode $\omega = 0$. Thus, there are a total of $N_{\text{AMC}} + 1$ AMC modes, and the set of AMC modes is represented as $\mathcal{S}_C := \{0, 1, 2, \ldots, N_{\text{AMC}}\}$. The channel conditions of the PRBs are assumed to make transitions only to the neighboring SNR intervals. This allows us to model the AMC mode transitions using an FSMC channel model [WM95], and the one-step transition probability from AMC mode ω to ω' is,

$$p_{\omega,\omega'} = 0, \qquad \forall |\omega' - \omega| > 1. \tag{7.3}$$

These one-step transition probabilities depend on the SNR thresholds and are given by the level crossing rates (LCRs) [WM95]. The LCR, L_ω, in the positive or negative going direction at an SNR level γ^b_ω of a fading signal envelope that follows Nakagami-m distribution, for any average SNR $\overline{\gamma^u}$, can be written as [LZG05],

$$L_\omega = \sqrt{2\pi \frac{m\gamma^b_\omega}{\bar{\gamma}}} \frac{f_m}{\Gamma(m)} \left(\frac{m\gamma^b_\omega}{\bar{\gamma}}\right)^{m-1} \exp\left(-\frac{m\gamma^b_\omega}{\bar{\gamma}}\right). \tag{7.4}$$

Here, f_m is the Doppler spread. The next state transition probability can be written as [RLM02, (5),(6)],

$$p_{\omega,\omega+1} = \frac{L_{\omega+1}\tau}{P_\omega}, \qquad \text{if } \omega = 0, 1, \ldots, N_{\text{AMC}} - 1, \qquad (7.5)$$

$$p_{\omega,\omega-1} = \frac{L_\omega \tau}{P_\omega}, \qquad \text{if } \omega = 1, \ldots, N_{\text{AMC}}, \qquad (7.6)$$

where P_ω, the probability that the channel is in state ω, is

$$P_\omega = \int_{\gamma_\omega^b}^{\gamma_{\omega+1}^b} p_\gamma(\gamma) d\gamma. \qquad (7.7)$$

For Nakagami-m distribution, (7.7) becomes [LZG05],

$$P_\omega = \frac{\Gamma\left(m, \frac{m\gamma_\omega^b}{\bar{\gamma}}\right) - \Gamma\left(m, \frac{m\gamma_{\omega+1}^b}{\bar{\gamma}}\right)}{\Gamma(m)}, \qquad (7.8)$$

where $\Gamma(m, \alpha) = \int_\alpha^\infty t^{m-1} e^{-t} dt$ is the complementary incomplete Gamma function. Using (7.5) and (7.6), the probabilities of transition to the same state are:

$$p_{\omega,\omega} = \begin{cases} 1 - p_{\omega,\omega+1} - p_{\omega,\omega-1}, & \text{if } 0 < \omega < N_{\text{AMC}} \\ 1 - p_{0,1}, & \text{if } \omega = 0, \\ 1 - p_{N_{\text{AMC}},N_{\text{AMC}}-1}, & \text{if } \omega = N_{\text{AMC}}. \end{cases} \qquad (7.9)$$

Equations (7.5), (7.6), and (7.9) constitute the elements of the one-step channel transition probability matrix $\vec{P}_{\text{ch}}$. One channel matrix is generated for each class using the average SNR of its tagged user.

7.1.2 Modeling the Medium Access Control (MAC) Layer: DP Scheduler

The behavior of the DP scheduler is governed by a $1 \times K$ control parameter vector $\vec{\xi} = [\xi_1, \xi_2, \cdots, \xi_K]$, such that $0 < \xi_k \leq 1$ and $\sum_{k=1}^K \xi_k = 1$. $\xi_k's$ are used to determine the order of priority in which the buffers $\mathcal{B}_k$ of the K classes are served. Thus, if $K = 3$ and $\vec{\xi} = [0.1, 0.55, 0.35]$, then it is observed after a very large observation time that the highest service priority has been enjoyed by class-*1*, class-*2*, and class-*3* in 10%, 55%, and 35% of the TTIs, respectively. The lower priority classes in any TTI are served using the PRBs which are left unused by the higher priority ones. Consequently, the service processes of the classes are coupled with one another. In the steady state, $\vec{\xi}$ gives the average resource/PRB share of each class.

Based on assumption A3, once a packet is assigned to a PRB, it has to be transmitted before the next TTI begins. So, this system has deterministic packet service times of the same duration as a TTI. Each resource/PRB can be mapped to a server and the user classes are the SNR regions/classes. The system thus translates into a finite buffer multi-server multi-class queueing model with deterministic service times. As in [LZG05, PRB10, RHHB09], and [LHA06], in this work, AMC results in an adaptive queue service process which causes a different number of packets to be transmitted per scheduling interval. However, in contrast to [LZG05, PRB10, RHHB09], and [LHA06], the queue service processes of the user classes are also coupled with each other. In the next section, queueing analysis is used to develop the analytical framework of the DP scheduler.

7.2 Queueing Analysis

In this section, considering Poisson packet arrivals, an augmented FSMC, which combines queueing with the DP scheduler behavior, and AMC, is derived.

7.2.1 Arrival Statistics

Let $A_k(t)$ denote the cumulative packet arrival process of class-k in the interval (0,t]. Since arrivals are Poisson, the probability that "a" number of class-k packets arrive in a TTI of duration τ is given by,

$$P\left(A_k(\tau) = a\right) = \frac{(\lambda_{p_k}\tau)^a \, e^{-\lambda_{p_k}\tau}}{a!}, \quad a \geq 0 \qquad (7.10)$$

where $\mathcal{E}\left\{A_k(\tau)\right\} = \lambda_{p_k}\tau$ [Kle75] is the number of packet arrivals per TTI. The process $\{A_k(\tau)\}$ is a stationary process. It is independent of the arrivals in other intervals [Kle75], the states of the queues, and of the PRBs. As the average service time is τ (TTI duration), the class-k traffic intensity is $\rho_k = \lambda_{p_k}\tau$, such that $\sum_{k=1}^{K}\rho_k \leq N_{\text{PRB}}$.

Packets arriving for class-k are stored in the finite-length buffer "$\mathcal{B}_k$" at the BS as long as it does not overflow. Buffered packets are accumulated in

$\mathcal{B}_k$ and served only at the beginning of TTIs according to the queue service process explained in Section 7.2.2.1. The state of "$\mathcal{B}_k$" at the beginning of any TTI "n" is denoted by its instantaneous length (in terms of the number of buffered packets) $b_{k,n-1} \in \mathcal{S}_{\mathcal{B}_k} := \{0, 1, \dots, L_k\}$.

As mentioned earlier, the state definition of the system should ideally include the buffer states and channel conditions (AMC modes) of all users in all PRBs as well as their order of service. Hence, even if a user is assumed to have the same channel condition across all PRBs, the introduction of every new user will increase the number of state variables by three (buffer states, channel conditions, and service order). This may render the system state space unscalable with the increase in the number of users. In contrast, in this model, it is assumed that all users of a class have the same buffer and channel conditions as the tagged user. So, the addition of a new user does not introduce any new state variable but only brings changes in the channel states of the tagged user of its class. This significantly reduces the state space. Nevertheless, real-life systems may need as many as 18 SNR-based user classes [BD15]. In those cases, analytical approximations or numerical methods are needed to solve the Markov chains [CLCPRAHV14]. Hence, for reasons of simplicity, the discussion in the rest of the chapter considers $K = 2$ classes [CLCPRAHV14].

7.2.2 Queue Service Process: DP Scheduler

The queue service process is dictated by the scheduling algorithm and the packet departure process.

7.2.2.1 Scheduler dynamics

We have discussed earlier that the behavior of the DP scheduler is governed by the scheduler control parameter, $\vec{\xi}$. Since there are $K = 2$ user classes, the scheduler control parameter can be represented as a scalar ξ. Let $\mathcal{S}_{\mathcal{K}} = \{1, 2\}$ represent the set of user classes such that users with $\overline{\gamma^u} \geq \gamma^{\text{th}}$ belong to the center-region and those with $\overline{\gamma^u} < \gamma^{\text{th}}$ belong to the edge-region. At the beginning of any TTI "n," the DP scheduler selects either class-*1* with probability ξ or class-*2* with probability $(1 - \xi)$. We represent the class which is assigned a higher priority in TTI "n" as $sc_n \in \mathcal{S}_{\mathcal{K}}$. Figure 7.2 shows the system model with $K = 2$ classes and one user in each class.

Let the prioritized class in any TTI "n" be class-k. So, the scheduler first serves the buffer "$\mathcal{B}_k$," i.e., it assigns PRBs to class-k according to the instantaneous length $b_{k,n-1}$ of "$\mathcal{B}_k$" and the AMC mode of $\mathbb{U}_k$. The

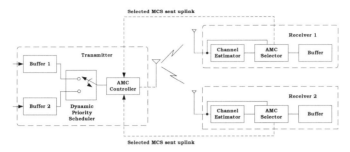

Figure 7.2 System model of the DP scheduler considering K=2 classes and only one user in each class.

remaining PRBs are assigned to the other class, j, $j \neq k$. Hence, unlike in [VDZF15] which models a strict priority non-pre-emptive scheduler, the DP scheduler can be used to model a wide range of schedulers. For example, when $\xi = 1$, the scheduler behaves as a strict priority scheduler for class-*1*. Since class-*1* represents the center-region users, $\xi = 1$ maximizes the sum-cell throughput, and thus the DP scheduler behaves as a greedy max-SINR scheduler [WMMZ11]. In the same way, when $\xi = 0$, class-*2* gets a higher priority. This increases the class-*2* throughput but reduces the overall throughput. The value of ξ can also be varied to obtain a weighted equal throughput for both classes. This value of ξ thus makes the scheduler behave as a fair scheduler.

7.2.2.2 Packet departure process

In this work, the packet departure process from the queues depends on the instantaneous PRB requirements of the classes. Let the number of PRBs required when the AMC mode is ω be denoted by R_ω. As the PRBs are mapped as servers, the AMC modes represent the server states. We define the server state or AMC mode of class-k in TTI "n" as $C_{k,n} \in \mathcal{S}_C := \{0, 1, 2, \ldots, N_{\mathrm{AMC}}\}$. It is to be noted that when AMC mode $C_{k,n} = 0$, the PRB requirement $R_{C_{k,n}} = \infty$. Since this gives rise to an impractical situation, no packets are transmitted when $C_{k,n} = 0, \forall k$.

We focus on any single TTI "n." Let class-k, for any $k \in \mathcal{S}_K$, have a higher priority of transmission in this TTI, and therefore, it has all N_{PRB} PRBs available to it. Thus, the maximum number of class-k packets transmitted/served during this TTI is $\left\lfloor \frac{N_{\mathrm{PRB}}}{R_{C_{k,n}}} \right\rfloor$. If the instantaneous buffer length

$b_{k,n-1} \geq \left\lfloor \frac{N_{PRB}}{R_{C_{k,n}}} \right\rfloor$, then the number of class-k packets served is $\phi_{k,n} = \left\lfloor \frac{N_{PRB}}{R_{C_{k,n}}} \right\rfloor$. Otherwise, when $b_{k,n-1} < \left\lfloor \frac{N_{PRB}}{R_{C_{k,n}}} \right\rfloor$, $\phi_{k,n} = b_{k,n-1}$. Therefore, the possible number of class-k packets that are assigned to the servers at the beginning of TTI "n" is,

$$\phi_{k,n} = \min \left(b_{k,n-1}, \left\lfloor \frac{N_{PRB}}{R_{C_{k,n}}} \right\rfloor \right). \tag{7.11}$$

The number of PRBs available to the other class-j (where $j \neq k, \forall k, j \in S_K$) is $N_{PRB} - \phi_{k,n} R_{C_{k,n}}$. Applying the same logic as for class-k, the number of the lower priority class-j packets transmitted in TTI "n" is,

$$\phi_{j,n} = \min \left(b_{j,n-1}, \left\lfloor \frac{N_{PRB} - \phi_{k,n} R_{C_{k,n}}}{R_{C_{j,n}}} \right\rfloor \right), \tag{7.12}$$

where $C_{k,n}$ may or may not be the same as $C_{j,n}$. We denote the number of PRBs available to class-k in any TTI n as $N_{PRB,k}^n$. It depends on the prioritized class, sc_n, and the instantaneous AMC modes of the two classes. Thus, when class-1 has a higher priority, $N_{PRB,1}^n = N_{PRB}$ and $N_{PRB,2}^n = N_{PRB} - \phi_{1,n} R_{C_{1,n}}$. Similarly, if class-$2$ has a higher priority, then $N_{PRB,2}^n = N_{PRB}$ and $N_{PRB,1}^n = N_{PRB} - \phi_{2,n} R_{C_{2,n}}$.

7.2.3 Synthesis of the Markov Chain

For generating the augmented FSMC for the queueing system discussed above, one needs to first develop the queue state recursion. Based on the queue service process discussed above, and its timeline in Figure 7.3, the queue state recursion can be written as,

$$\begin{aligned} b_{k,n} &= \min \left(L_k, \mathcal{L}_{k,n} + \mathcal{A}_{k,n} \right) \\ &= \min \left(L_k, (b_{k,n-1} - \phi_{k,n}) + \mathcal{A}_{k,n} \right). \end{aligned} \tag{7.13}$$

From (7.11)–(7.13) and Figure 7.3, it is observed that the buffer states $b_{k,n}$, $\forall k$, at the beginning of TTI "$n+1$" depend on the buffer states $b_{k,n-1}$, the server states $C_{k,n}$, the prioritized class sc_n, and the arrival process in the previous TTI "n." As the arrivals are Poisson, the system can be modeled as an *imbedded Markov chain* [Kle75] whose states are observed at the beginning of every TTI immediately before the buffers are served. Moreover,

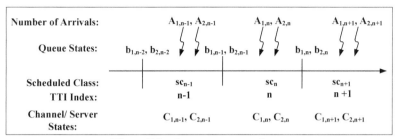

Figure 7.3 Queue state recursion.

as the buffers are of finite length and there are a finite number of AMC modes, the state space of the system is a discrete time FSMC, which is an augmented form of the Markov chain of the channel, given by (7.5), (7.6), and (7.9). Thus, the state space of the DP scheduler for two user classes can be defined as follows.

Definition 7.2.1. *In any TTI "n," the buffer state $b_{k,n-1}$ and the server state $C_{k,n}$ of class-k, $\forall k \in S_K$, and the prioritized scheduler class sc_n are independent of each other. So, the state space of the DP scheduler is,*

$$S_{DP} := \{(\mathbf{b_{n-1}}, sc_n, \mathbf{C_n})\},\tag{7.14}$$

where $\mathbf{b_{n-1}} = (b_{1,n-1}, b_{2,n-1}) \in S_{B_1} \times S_{B_2}$ denotes the buffer states, $\mathbf{C_n} = (C_{1,n}, C_{2,n}) \in S_C \times S_C$ represents the server states, and $sc_n \in S_K$ represents the prioritized class of TTI "n."

The number of states in S_{DP} is $K \prod_{i=1}^{K} (L_k + 1) (N_{AMC} + 1)^K$. The QoS metrics supported by the DP scheduler can be obtained from the stationary distribution of the augmented FSMC $(b_{1,n-1}, b_{2,n-1}, sc_n, C_{1,n}, C_{2,n})$ of (7.14). For this, the stationary state distribution should exist and should be unique. The proof is available in [Pal18]. Let the stationary state distribution be given by,

$$\pi^{DP}_{(b_1=x,b_2=y,sc=k,C_1=\omega_1,C_2=\omega_2)}$$
$$= \lim_{n\to\infty} P(b_{1,n-1} = x, b_{2,n-1} = y, sc_n = kC_{1,n} = \omega_1, C_{2,n} = \omega_2)\tag{7.15}$$

where π^{DP}_s denotes the steady-state probability of any state s of the FSMC of (7.14). The steady-state probability vector is $\vec{\pi}^{DP} = [\pi^{DP}_0, \pi^{DP}_1, \ldots, \pi^{DP}_{|S_{DP}|}]$

and can be obtained by solving [Kle75],

$$\pi^{\vec{\mathrm{DP}}} = \pi^{\vec{\mathrm{DP}}}\vec{\mathbf{P}}, \qquad \sum_{s=1}^{|\mathcal{S}_{\mathrm{DP}}|} \pi_s^{\mathrm{DP}} = 1, \qquad (7.16)$$

where $\vec{\mathbf{P}}$ is the one-step transition probability matrix. According to Lemma 2 of the Appendix, the FSMC of (7.14) is homogeneous in nature. This implies that the probability of transition from state $s_n = (b_{1,n-1} = x, b_{2,n-1} = y, sc_n = k, C_{1,n} = \omega_1, C_{2,n} = \omega_2)$ in TTI "n" to state $s_{n+1} = (b_{1,n} = x', b_{2,n} = y', sc_{n+1} = k', C_{1,n+1} = \omega_1', C_{2,n+1} = \omega_2')$ in TTI "$n + 1$" is independent of time. This state transition probability when $sc_n = 1$ is:

$$
\begin{aligned}
P_{s_n \to s_{n+1}} &= P\left(s_{n+1}|s_n\right) \\
&= P\left(b_{1,n} = x'|b_{1,n-1} = x, sc_n = 1, C_{1,n} = \omega_1\right) \\
&\quad \times P(b_{2,n} = y'|b_{2,n-1} = y, C_{2,n} = \omega_2, sc_n = 1, \\
&\qquad b_{1,n-1} = x, C_{1,n} = \omega_1) \\
&\quad \times P\left(C_{1,n+1} = \omega_1'|C_{1,n} = \omega_1\right) \times P\left(C_{2,n+1} = \omega_2'|C_{2,n} = \omega_2\right) \\
&\quad \times P\left(sc_{n+1} = k'\right).
\end{aligned} \qquad (7.17)
$$

Since class-*1* has a higher priority in TTI "n," the number of PRBs available to it is $\mathrm{N}_{\mathrm{PRB},1}^n = \mathrm{N}_{\mathrm{PRB}}$. Consequently, $\phi_{1,n} = \min\left(x, \left\lfloor \frac{\mathrm{N}_{\mathrm{PRB}}}{R_{\omega_1}} \right\rfloor\right)$ gives the number of class-*1* packets served. The number of packets left in $\mathcal{B}_1$ after serving $\phi_{1,n}$ packets is $\left(x - \min\left(x, \left\lfloor \frac{\mathrm{N}_{\mathrm{PRB}}}{R_{\omega_1}} \right\rfloor\right)\right) = \max\left(0, \left(x - \left\lfloor \frac{\mathrm{N}_{\mathrm{PRB}}}{R_{\omega_1}} \right\rfloor\right)\right)$. So, the probability that there are x packets in $\mathcal{B}_1$ at TTI "n" and x' packets at TTI "$n + 1$" is given by the probability of arrival of $\mathcal{A}_{1,n} = x' - \max\left(0, x - \left\lfloor \frac{\mathrm{N}_{\mathrm{PRB}}}{R_{\omega_1}} \right\rfloor\right)$ packets. Thus, the first term of the last equality of (7.17) can be expressed as,

$$
\begin{aligned}
&P\left(b_{1,n} = x'|b_{1,n-1} = x, sc_n = 1, C_{1,n} = \omega_1\right) \\
&= \begin{cases} P\left(\mathcal{A}_{1,n} = x' - \max\left(0, x - \left\lfloor \frac{\mathrm{N}_{\mathrm{PRB}}}{R_{\omega_1}} \right\rfloor\right)\right), \\ \qquad\qquad\qquad\qquad\qquad \text{when } 0 \le x' < L_1, \\ 1 - \sum_{0 \le x' < L_1} P\left(b_{1,n} = x'|b_{1,n-1} = x, sc_n = 1, C_{1,n} = \omega_1\right), \\ \qquad\qquad\qquad\qquad\qquad \text{when } x' = L_1. \end{cases}
\end{aligned} \qquad (7.18)
$$

In the same TTI, the number of PRBs available to class-2 is $\mathrm{N}^{n}_{\mathrm{PRB},2} = \mathrm{N}_{\mathrm{PRB}} - \phi_{1,n} R_{\omega_1}$. So, applying the same logic as class-1, the second term of the last equality of (7.17) can be expressed as:

$$P\left(b_{2,n} = y'|b_{2,n-1} = y, C_{2,n} = \omega_2, sc_n = 1, b_{1,n-1} = x, C_{1,n} = \omega_1\right)$$

$$= \begin{cases} P(A_{2,n} = y' - \max & \text{when } 0 \le y' < L_2 \\ (0, y - \left\lfloor \frac{\mathrm{N}_{\mathrm{PRB}} - \phi_{1,n} R_{\omega_1}}{R_{\omega_2}} \right\rfloor)), & \\ 1 - \sum\limits_{0 \le y' < L_2} P(b_{2,n} = y'|b_{2,n-1} = y, & \text{when } y' = L_2. \\ C_{2,n} = \omega_2, sc_n = 1, b_{1,n-1} = x, C_{1,n} = \omega_1), & \end{cases}$$

$$(7.19)$$

The values of $P\left(C_{1,n+1} = \omega'_1|C_{1,n} = \omega_1\right)$ and $P\left(C_{2,n+1} = \omega'_2|C_{2,n} = \omega_2\right)$ in (7.17) are obtained from the transition probabilities of the FSMC channel model in (7.5), (7.6), and (7.9). Finally, $P\left(sc_{n+1} = k'\right) = \xi$ when $k' = 1$ and $P\left(sc_{n+1} = k'\right) = (1 - \xi)$ when $k' = 2$. In the same way, the elements of $\vec{P}$ can also be derived when $sc_n = 2$.

7.3 Evaluation of Performance Metrics

In this section, the steady-state probabilities π^{DP}_s derived from (7.16) are used to obtain some important QoS metrics.

7.3.1 Packet Drop Probability

Packet drops occur due to buffer overflow. Let the number of class-k packets dropped in TTI "n" be denoted by $\Psi_{k,n}$. The number of free slots in the buffer $\mathcal{B}_k$ in this TTI after $\phi_{k,n}$ packets are served is given by $\alpha_{k,n} = \mathcal{A}_{k,n} - \left(L_k - \max\left(0, b_{k,n-1} - \left\lfloor \frac{\mathrm{N}^{n}_{\mathrm{PRB},k}}{R_{C_{k,n}}} \right\rfloor\right)\right)$. If the number of packets arriving to $\mathcal{B}_k$, i.e., $\mathcal{A}_{k,n}$, is greater than $\alpha_{k,n}$, then $(\mathcal{A}_{k,n} - \alpha_{k,n})$ packets are dropped; otherwise, no packet is dropped. Thus, $\Psi_{k,n}$ can be expressed as,

$$\Psi_{k,n} = \max\left(0, \mathcal{A}_{k,n} - \alpha_{k,n}\right)$$

$$= \max\left(0, \mathcal{A}_{k,n} - \left(L_k - \max\left(0, b_{k,n-1} - \left\lfloor \frac{\mathrm{N}^{n}_{\mathrm{PRB},k}}{R_{C_{k,n}}} \right\rfloor\right)\right)\right)$$

$$(7.20)$$

The stationary distribution of the FSMC in (7.14) exists and is unique. Let (b_1, b_2, sc, C_1, C_2) be a state of this Markov chain in the steady state. Moreover, $A_{k,n}$ is also a stationary process, such that $A_k = \lim_{n \to \infty} A_{k,n}$. Thus, $\forall k \in \mathcal{S}_K$ the stationary distribution of $\Psi_{k,n}$ is given by,

$$\Psi_k = \lim_{n \to \infty} \Psi_{k,n} = \max \left(0, A_k - \left(L_k - \max \left(0, b_k - \left\lfloor \frac{N_{\text{PRB},k}^n}{R_{C_k}} \right\rfloor \right) \right) \right). \tag{7.21}$$

It is to be noted that as $R_{C_k} = \infty$ when $C_k = 0$, (7.21) can be applied only when $C_k \neq 0$. As no packets are served for $C_k = 0$, the number of class-k packets dropped is,

$$\Psi_k = \max \left(0, A_k - L_k + b_k \right). \tag{7.22}$$

The probability of dropping class-k packets, P_{D_k}, is obtained as the ratio of the average number of packets dropped to the average number of arrivals in a TTI [LZG05] such that,

$$P_{D_k} = \lim_{T \to \infty} \frac{\sum\limits_{n=1}^{T} \Psi_{k,n}}{\sum\limits_{n=1}^{T} A_{k,n}} = \frac{\mathcal{E}\{\Psi_k\}}{\lambda_{p_k} T}. \tag{7.23}$$

The average number of class-*1* packets dropped can be derived as

$$\mathcal{E}\{\Psi_1\} = \sum_{b_1=0}^{L_1} \sum_{b_2=0}^{L_2} \sum_{k=1}^{2} \sum_{C_1=0}^{N_{\text{AMC}}} \sum_{C_2=0}^{N_{\text{AMC}}} \pi_{(b_1,b_2,1,C_1,C_2)}^{\text{DP}}$$

$$\times \sum_{u=0}^{\infty} u Pr \left(A_1 = u + L_1 - \max \left(0, b_1 - \left\lfloor \frac{N_{\text{PRB}}}{R_{C_1}} \right\rfloor \right) \right)$$

$$+ \sum_{b_1=0}^{L_1} \sum_{b_2=0}^{L_2} \sum_{C_1=1}^{N_{\text{AMC}}} \sum_{C_2=1}^{N_{\text{AMC}}} \pi_{(b_1,b_2,2,C_1,C_2)}^{\text{DP}}$$

$$\times \sum_{u=0}^{\infty} u Pr \left(A_1 = u + L_1 - \max \left(0, b_1 - \left\lfloor \frac{N_{\text{PRB}} - \phi_2 R_{C_2}}{R_{C_1}} \right\rfloor \right) \right) \tag{7.24}$$

The first term in (7.24) gives the average number of class-*1* packets dropped when class-*1* has a higher priority. The second term corresponds to the case when class-*2* has a higher priority. In this case, as the service process of class-*2* affects that of class-*1*, the summation is carried out only over the non-zero values of C_2 as otherwise the number of resources required by class-*2* would be infinite. Using the same logic, one can find out the average number of class-*2* packets dropped [Pal18].

7.3.2 Throughput

Using the probability of packet drop derived earlier, one can find the class-k throughput, which depends on the number of correctly received packets. Packets can be received correctly only if they enter the buffer $\mathcal{B}_k$. The number of class-k packets entering $\mathcal{B}_k$ is $\lambda_{p_k}(1 - P_{D_k})$. The probability of receiving these packets correctly is $(1 - P_e)$, where P_e is the packet error rate. In this work, it is assumed that a block of data consists of a single packet. Hence, the packet error rate is the BLER. Thus, the average class-k throughput is:

$$\mathbb{T}_k = \lambda_{p_k}\left(1 - P_{D_k}\right)(1 - P_e).\tag{7.25}$$

7.3.3 Average Delay

Apart from the probability of packet drop and throughput, the average delay of a class-k packet is also a very important metric and is given by:

$$\delta_k = \delta_{Q_k} + \delta_{\text{TTI}},\tag{7.26}$$

where δ_{Q_k} is the average queueing delay of a class-k packet and $\delta_{\text{TTI}} = \frac{\tau}{2}$ is the delay caused due to serving packets at the beginning of every TTI. The queueing delay of a class-k packet can be obtained using Little's law [Kle75] as:

$$\delta_{Q_k} = \frac{\overline{\mathcal{B}_k}}{(1 - P_{D_k})\lambda_{p_k}},\tag{7.27}$$

where $\overline{\mathcal{B}_k}$ the average queue length of class-k is:

$$\overline{\mathcal{B}_k} = \sum_{b_1=0}^{L_1}\sum_{b_2=0}^{L_2}\sum_{k=1}^{2}\sum_{C_1=0}^{N_{\text{AMC}}}\sum_{C_2=0}^{N_{\text{AMC}}}\pi_{(b_1,b_2,sc,C_1,C_2)}^{\text{DP}}b_k.\tag{7.28}$$

7.4 Results

7.4.1 Evaluation Framework

The average SNRs of the tagged user of the center and edge regions are taken to be $\gamma_1 = 20$dB and $\gamma_2 = 10$dB, respectively. The Nakagami fading parameter m is set to 1, implying a Rayleigh fading channel. The simulation parameters outlined in Table 7.2 are largely adopted from Third Generation Partnership Project (3GPP) LTE [IR08]. Although the model is generalized for any number of PRBs/servers and AMC modes, for time and computational resource limitations, the numbers of PRBs and AMC modes are taken as $N_{PRB} = 15$ and $N_{AMC} = 3$. The AMC modes for a BLER of 0.1 are given in Table 7.3 [STB09]. The same table also gives the number of servers required by SISO systems in each AMC mode to transmit a packet of $N_b = 320$ bits (a voice packet in adaptive multi-rate 12.2 Kbps VoIP codec is of 40 bytes or 320 bits). To average out the packet delay in case of simulations, the length of the buffers is 1.5 times of N_{PRB}. So, $L_k = (1.5 N_{PRB}) = 23, \forall k$.

7.4.2 Verification of Performance Analysis

In this section, discrete event simulation results used for validating the analytical framework of the DP scheduler are presented. Both the analytical and

Table 7.2   Simulation parameters

Parameters	Values
Carrier frequency	2 GHz
Number of PRBs or servers	15
TTI duration	1 ms
Buffer length	23
Doppler frequency, f_m	10 Hz
Number of subcarriers per PRB	12
Number of symbols per PRB	14
Number of control symbols per PRB	3
Number of pilot REs per PRB	8

Table 7.3   AMC modes used, BLER=0.1 for all modes [STB09, Table 10.1]

Modulation(bits/symbol)	QPSK	16-QAM	64-QAM
Coding rate	1/2	1/2	1/2
R_n (bits/symbol)	1	2	3
Number of PRBs required	3	2	1
SNR thresholds	-1.2	6.2	11.3

QAM stands for quadrature amplitude modulation.

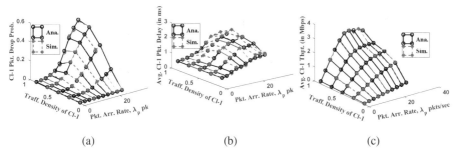

Figure 7.4   Variation with average arrival rate and traffic density of (a) packet drop probability of class-*1*, (b) average delay of class-*1* packets, and (c) class-*1* throughput with $\xi = 0.5$.

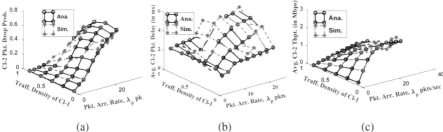

Figure 7.5   Variation with average arrival rate and traffic density of (a) packet drop probability of class-*2*, (b) average delay of class-*2* packets, and (c) class-*2* throughput with $\xi = 0.5$.

simulation results are generated with the scheduler control parameter $\xi = 0.5$ and five different traffic densities, $\zeta = [0.1\ 0.9]$, $\zeta = [0.3\ 0.7]$, $\zeta = [0.5\ 0.5]$, $\zeta = [0.7\ 0.3]$, and $\zeta = [0.9\ 0.1]$. For each ζ, 10 different values of the total packet arrival rate λ_p varying between $[0.5,\ 24]$ are considered and for each λ_p simulated, the arrival and departure of $100,000$ packets have been simulated.

Figures 7.4 and 7.5 show the variation of QoS metrics, such as average delay, packet drop probability, and average throughput, of both classes as a function of the average arrival rate and user density. It is seen that there is a close match between the simulation results and the analytical framework.

7.5 Call Admission Control Using the DP Scheduler Framework

The developed analytical frameworks can be used for estimating QoS metrics which are used for CAC. The aim is to maximize the number of admitted calls while guaranteeing multiple QoS constraints, such as average packet delay and packet drop probability, by adjusting the scheduler control parameter ξ. The QoS metrics are estimated using the proposed scheduler framework.

We use time-scale decomposition for implementing the call level and packet level dynamics [Hui88]. It is assumed that within the duration of observation T_{obs}, each user in a cell of area $\mathcal{C}_\mathcal{R}$ can generate only one call and each has a different average SNR $\overline{\gamma^u}$. Call arrivals in $\mathcal{C}_\mathcal{R}$ follow a Poisson process of rate λ_c and have an exponentially distributed call duration with mean $\frac{1}{\mu_c}$. Based on $\overline{\gamma^u}$, the calls may be classified into center and edge regions. Thus, call arrival rate in region/class-k is $\lambda_c \zeta_k$, where ζ_k as before is the traffic share of class-k. For each call, packets arrive according to a Poisson process of rate λ_p packets/s. Therefore, when there are N_c^{AD} ongoing calls or sessions in the system, the total packet arrival rate is $\lambda_p \cdot N_c^{\mathrm{AD}}$ packets/s.

7.5.1 Working of the CAC and Role of the Scheduler

The flowchart of the CAC procedure is given in Figure 7.6. We observe the system for all calls arriving between $t_{\mathrm{now}} = 0$ to $t_{\mathrm{now}} = T_{\mathrm{obs}}$. Thus, the required inputs are:

1) the observation interval T_{obs}, 2) the packet delay threshold δ_{th}, and 3) the packet drop probability threshold $P_{D_{th}}$. For each call, the system takes as input: 1) the existing class-wise average SNR, $\overline{\gamma_k}$, $\forall k$, and 2) the current scheduler control parameter ξ_{curr}. Once a call arrives from a user "u_{new}" with average SNR $\overline{\gamma^{u_{new}}}$, the CAC module places it in class-*1* if $\overline{\gamma^{u_{new}}} \geq \gamma^{th}$ or in class-*2* if $\overline{\gamma^{u_{new}}} < \gamma^{th}$. The temporary average SNR, $\overline{\gamma_{k,new}}$, of class-$k$ to which the new user is sorted is,

$$\overline{\gamma_{k,new}} = \frac{1}{N_c^{\mathrm{AD}} \zeta_k + 1} \left(\overline{\gamma_k} N_c^{\mathrm{AD}} \zeta_k + \overline{\gamma^{u_{new}}} \right). \tag{7.29}$$

The CAC unit next runs the mathematical framework of the DP scheduler of Section 7.2.2. It uses the updated class-wise SNR to generate the FSMC of the channel for each class using (7.5), (7.6), and (7.9). With the updated channel matrices and $\xi_{\mathrm{mod}} = \xi_{\mathrm{curr}}$, it solves the scheduler framework using (7.3)–(7.18) and then obtains the class-wise average packet delay δ_k from (7.26) and packet drop probability P_{D_k}, $\forall k$ from (7.23). These QoS

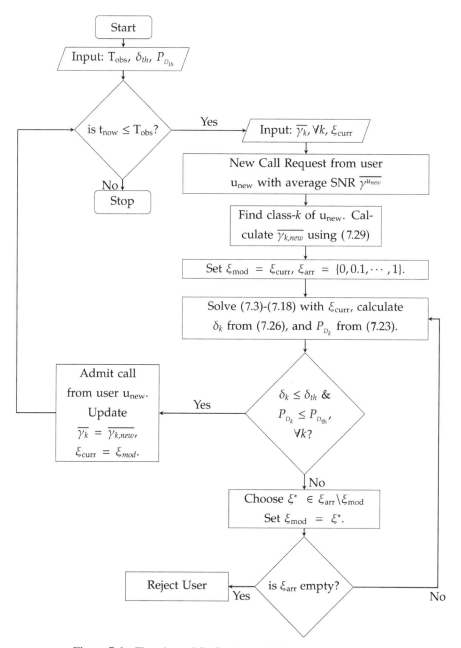

Figure 7.6 Flowchart of CAC using the DP scheduler framework.

metrics, which are predicted under the assumption that the new call is already admitted, are compared with the predefined delay and packet drop probability thresholds, δ_{th} and $P_{D_{th}}$, respectively. If both δ_k and P_{D_k} satisfy the following constraint in (7.30), only then the call is admitted.

$$\delta_k \leq \delta_{th}, \qquad P_{D_k} \leq P_{D_{th}}. \qquad (7.30)$$

If, however, the constraint is not satisfied, then the CAC unit exhaustively searches a $\xi^* \in \xi_{arr} \backslash \xi_{mod}$ such that the constraint in (7.30) is satisfied and the call gets admitted. Here, ξ_{arr} represents the set of all possible values of the scheduler control parameter. If no such ξ^* exists, then the call is rejected. Otherwise, the call is admitted and $\overline{\gamma_k}$, ξ_{curr} are updated. By adjusting ξ_{curr}, the resource share of the two classes is adjusted so that the arriving call can be accommodated. Packets of the admitted calls are scheduled using the DP scheduler. They are scheduled to the PRBs only at the beginning of every TTI, as seen in Figure 7.3. However, arriving calls can be admitted at any point of time. It has been demonstrated in Figure ?? that the QoS metrics depends on ξ. Thus, for a given user density and QoS constraint, ξ can be controlled to increase the system's call carrying capacity.

7.5.1.1 Simulation setup

The results of CAC are carried out for 1000 calls. Each call comes from a different user. 10,000 users are uniformly distributed within a circular cell of radius $R = 2000$ m and their average SNR are calculated according to Section ??. Of these 10,000 users, 1000 users are chosen randomly. The number of PRBs is $N_{PRB} = 6$. The call arrival rate is varied between [0.5, 6]. Each call has an average duration of $h_c = 500$ ms and a packet arrival rate of $\lambda_p = 100$ packets/s. This translates to a total of $P_n = 50,000$ packets for each call on an average. Other parameters remain the same as Table ??. The AMC modes, packets sizes, and SNR threshold for user classification remains the same as in earlier sections.

7.5.1.2 Results on CAC for the DP scheduler framework

The fraction of calls generated by class-k is taken to be $\zeta_k = 0.5, \forall k$. Since there is no closed-form expression relating ξ with the key performance indicators, extensive discrete event simulations are used to obtain the value of the scheduler control parameter ξ which admits a call while satisfying a given delay and packet drop probability constraint. Results are compared with those of no admission control.

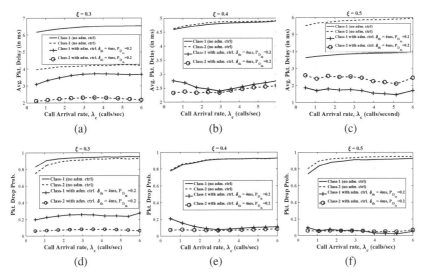

Figure 7.7 Average packet delay of admitted calls when $\zeta = [0.5\ 0.5]$ for (a) $\xi = 0.3$, (b) $\xi = 0.4$, and (c) $\xi = 0.5$. Probability of packet Drop of admitted calls when $\zeta = [0.5\ 0.5]$ for (d) $\xi = 0.3$, (e) $\xi = 0.4$, and (f) $\xi = 0.5$.

Figures 7.7(a–c) and (d–f), respectively, show the average delay and packet drop probability of the two classes for different values of ξ, and packet delay threshold $\delta_{th} = 4$ ms and packet drop probability threshold $\psi_{th} = 0.2$. These values are solely for the purpose of demonstrating the role of the developed analytical framework in CAC.

It is seen from Figures 7.7(a–c) that the average packet delay constraint is satisfied for all the selected values of ξ. So, the choice of ξ depends wholly on the packet drop probability. When $\xi = 0.3$, the packet drop probability constraint is seen to be satisfied only for $\lambda_c = 0.5$ calls/s. While it is seen from Figure 7.7(e) that for $\xi = 0.4$, the constraint is satisfied for values of λ_c greater than 0.5 calls/s. So, initially the algorithm starts with $\xi_{curr} = 0.3$. As the call arrival rate increases, the scheduler control parameter value has to be increased by the CAC unit to $\xi_{curr} = 0.4$ for the packet drop probability constraint to be satisfied. The scheduler control parameter determines the resource share of each class. So, with the increase in call arrival rate, ξ needs to be adjusted so as to satisfy the QoS constraints of (7.30). It is also observed from Figure 7.7(c) that with $\xi = 0.5$, both the QoS constraints in (7.30) are found to be satisfied for all classes at all call arrival rates. This implies that a resource share of $(1 - \xi) = 0.5$ for class-*2* is sufficient for providing its predefined QoS guarantees.

Table 7.4   Values of scheduler control parameter ξ, for different values of traffic density ζ for $\delta_{th} = 4$ ms and $P_{D_{th}} = 0.2$

ζ	[0.3 0.7]	[0.5 0.5]	[0.7 0.3]
ξ	0.3	0.4	0.6

Considering that the QoS requirements of $\delta_{th} = 4$ ms and $P_{D_{th}} = 0.2$ are stringent, the values of ξ which satisfy the QoS constraints for different traffic densities at all traffic arrival rates are given in Table 7.4. It may, therefore, be inferred from the results presented above that the DP scheduler framework can be used for CAC and its characterizing control parameter can be used to improve the call blocking probability of the system while satisfying predefined QoS constraints.

7.6 Summary

An analytical framework of a DP RRA algorithm serving multiple SNR-based user classes in the downlink of OFDMA wireless access networks is designed using queueing theory in the chapter above. The framework is easily applicable for multi-class QoS traffic scenario. The DP scheduler behavior has been characterized by a control parameter that determines the long-term resource share of each class. An augmented FSMC which combines queueing, scheduler behavior, and AMC has been developed to derive QoS metrics like average delay, packet drop probability, and throughput which is used for predictive CAC realization for maximization of the number of admitted calls. The developed analytical framework may be useful especially in small cell scenarios and indoor hotspots where a small number of users are attached to the BS. The framework may also find use in multicast satellite systems.

References

[20112] IEEE Standard 802.11ad 2012. *Part 11: Wireless LAN Medium Access Control (MAC) and Physical Layer (PHY) Specifications. Amendment 3: Enhancements for Very High Throughput in the 60 GHz Band.* Technical Report, IEEE, 2012.

[3gp03] 3rd Generation Partnership Project. *Technical Specification Group Radio Access Network; Spatial Channel Model for MIMO Simulations.* Technical Report, 3GPP, 2003.

[3gp05] Technical Specification Group Radio Access Network. *Home Node B Radio Frequency (RF) Requirements (FDD): Release 9.* Tech Report, 3GPP. TR 25.967 V9.0.0, 3rd Generation Partnership Project, 2009.

[3GP08] 3GPP. Telecommunication Management; Self-Organizing Networks (SON); Concepts and Requirements. TS 32.500, 3rd Generation Partnership Project (3GPP), 2008.

[3GP08] 3GPP. Evolved Universal Terrestrial Radio Access (E-UTRA); Physical Layer Procedures. TS 36.213, 2008.

[3GP10] 3GPP. *Further Advancements for E-UTRA Physical Layer Aspects.* Technical Report. TR 36.814 v2.0.10, 2010.

[3GP11] 3GPP. Telecommunication Management; Energy Saving Management (ESM); Concepts and Requirements. TS 32.551, 3rd Generation Partnership Project (3GPP), 2011.

[3GP12] 3GPP. Telecommunication Management; Self-Organizing Networks (SON) Policy Network Resource Model (NRM) Integration Reference Point (IRP). TS 32.521, 3rd Generation Partnership Project (3GPP), 2012.

[3gp15] 3rd Generation Partnership Project. *Technical Specification Group Radio Access Network; Study on 3D Channel Model for LTE.* Technical Report, 3GPP, 2015.

[3gp17a] ETSI TR 125 996 V13.1.0 (2017-01): Universal Mobile Telecommunications System (UMTS). *Spatial Channel Model for Multiple Input Multiple Output (MIMO) Simulations.* Technical Report, 3GPP, 2017.

[3gp17b] 3rd Generation Partnership Project. *Technical Specification Group Radio Access Network; Study on New Radio (NR) Access Technology;RF and Co-existence Aspects (release 14)*. Technical Report, 3GPP, 2017.

[3gp17c] 3rd Generation Partnership Project. *Technical Specification Group Radio Access Network; Study on New Radio (NR) Access Technology;Radio Interface Protocol Aspects (release 14)*. Technical Report, 3GPP, 2017.

[3GP18] 3GPP. *User Equipment Radio Transmission and Reception; Technical Specification Part 1: Range 1 Standalone (release 15)*. Technical Report, 3GPP, 2018.

[5Gs] Available at: http://www.5gsummit.org/.

[80209] *Part 15.3: Wireless Medium Access Control (MAC) and Physical Layer (PHY) Specifications for High Rate Wireless Personal Area Networks (WPANs), Amendment 2: Millimeter-Wave-Based Alternative Physical Layer Extension*. Technical Report, IEEE 802.15.3c, IEEE, 2009.

[A+07] P. Almers et al. Survey of Channel and Radio Propagation Models for Wireless MIMO Systems. *EURASIP Journal on Wireless Communications and Networking*, 2007.

[A+14] J. G. Andrews et al. What Will 5G Be? *IEEE Journal on Selected Areas in Communications*, 32(6):1065–1082, 2014.

[ABC⁺14] J. G. Andrews, S. Buzzi, Wan Choi, S. V. Hanly, A. Lozano, A. C. K. Soong, and J. C. Zhang. What Will 5G Be? *IEEE Journal on Selected Areas in Communications*, 32(6):1065–1082, 2014.

[ACD+12] J. G. Andrews, H. Claussen, M. Dohler, S. Rangan, and M. C. Reed. Femtocells: Past, Present, and Future. *IEEE Journal on Selected Areas in Communications*, 30(3):497–508, 2012.

[ADH+94] K. Allen, N. DeMinco, J. R. Hoffman, Y. Lo, and P. B. Papazian. Building Penetration Loss Measurements at 900 MHz, 11.4 GHz, and 28.8 GHz. *U.S. Dept. Commerce, National Telecommunications Inf. Admin. (NTIA), Boulder, CO, USA, Rep. 94–306*, 1994.

[Adh08] P. Adhikari. Understanding Millimeter Wave Wireless Communication. *Whitepaper from Loea Corporation, San Diego*, 2008.

[AGG⁺11] G. Auer, V. Giannini, I. Godor, P. Skillermark, M. Olsson, M. A. Imran, D. Sabella, M. J. Gonzalez, C. Desset, and O. Blume. Cellular Energy Efficiency Evaluation Framework. In *IEEE 73rd Vehicular Technology Conference (VTC Spring)*, 1–6, 2011.

[agi13] *Wireless LAN at 60 GHz - IEEE 802.11ad Explained*. Technical Report, Agilent, USA, 2013.

[Ahm05] M. H. Ahmed. Call admission Control in wireless Networks: A comprehensive Survey. In *IEEE Communications Surveys Tutorials*, 7(1):49–68, 2005.

[AIIE13] O. G. Aliu, A. Imran, M. A. Imran, and B. Evans. A Survey of Self Organisation in Future Cellular Networks. In *IEEE Communications Surveys and Tutorials*, 15(1):336–361, 2013.

[AJG05] A. Abraham, L. C. Jain, and R. Goldberg. *Evolutionary Multiobjective Optimization: Theoretical Advances and Applications*. Springer-Verlag New York, Inc., Secaucus, NJ, USA, 2005.

[AJM15] J. Abdoli, M. Jia, and J. Ma. Filtered OFDM: A New Waveform for Future Wireless Systems. In *IEEE 16th International Workshop on Signal Processing Advances in Wireless Communications (SPAWC)*, 66–70, 2015.

[ALS+13] M. R. Akdeniz, Y. Liu, M. K. Samimi, S. Sun, S. Rangan, and T. S. Rappaport. *Millimeter Wave Channel Modeling and Cellular Capacity Evaluation*. Available at: http://arxiv.org/abs/1312.4921, 2013.

[ALS+14] M. R. Akdeniz, Y. Liu, M. K. Samimi, S. Sun, S. Rangan, T. S. Rappaport, and E. Erkip. Millimeter Wave Channel Modeling and Cellular Capacity Evaluation. In *IEEE Journal on Selected Areas in Communications*, 32(6):1164–1179, 2014.

[AM13] A. Asadi and V. Mancuso. A Survey on Opportunistic Scheduling in Wireless Communications. In *IEEE Communications Surveys Tutorials*, 15(4):1671–1688, Fourth Quarter 2013.

[AM15] P. V. Amadori and C. Masouros. Low RF-Complexity Millimeter-Wave Beamspace-MIMO Systems by Beam Selection. In *IEEE Transactions on Communications*, 63(6):2212–2223, 2015.

[AMS13] H. Ajorloo and M. T. Manzuri-Shalmani. Modeling Beacon Period Length of the UWB and 60-GHz mmWave WPANs Based on ECMA-368 and ECMA-387 Standards. In *IEEE Transactions on Mobile Computing*, 12(6):1201–1213, 2013.

[AS72] M. Abramowitz and I. Stegun. *Handbook of Mathematical Functions with Formulas, Graphs, and Mathematical Tables*. Dover, 1972.

[AT05] L. An and P. Tao. Annals of Operations Research, 2005.

[ATR16a] F. Ademaj, M. Taranetz, and M. Rupp. 3GPP 3D MIMO Channel Model: A Holistic Implementation Guideline for Open Source Simulation Tools. *EURASIP Journal on Wireless Communications and Networking*, 2016(1):1, 2016.

[ATR16b] F. Ademaj, M. Taranetz, and M. Rupp. 3GPP 3D MIMO Channel Model: A Holistic Implementation Guideline for Open Source Simulation Tools. *EURASIP Journal on Wireless Communications and Networking*, 2016(1):1, 2016.

[AWW+13] Y. Azar, N. Wong, K. Wang, R. Mayzus, J. K. Schulz, H. Zhao, F. Gutierrez Jr., D. Hwang, and T. S. Rappaport. 28 GHz Propagation Measurements for Outdoor Cellular Communications Using Steerable Beam Antennas in New York City. In *IEEE International Conference on Communications (ICC)*, Budapest, Hungary, 2013.

[B+11] T. Baykas et al. IEEE 802.15.3c: The First IEEE Wireless Standard for Data Rates over 1 Gb/s. *IEEE Communications Magazine*, 114–121, 2011.

[BAB05] L. T. Bui, H. A. Abbass, and J. Branke. Multiobjective Optimization for Dynamic Environments. In *The IEEE Congress on Evolutionary Computation*, 3, 2349–2356, 2005.

[BAH14] T. Bai, A. Alkhateeb, and R. W. Heath Jr. Coverage and Capacity of Millimeter-Wave Cellular Networks. *IEEE Communication Magazine*, 71–78, 2014.

[BBS13] J. Brady, N. Behdad, and A. M. Sayeed. Beamspace MIMO for Millimeter-Wave Communications: System Architecture, Modeling, Analysis, and Measurements. *IEEE Transactions on Antennas and Propagation*, 61(7):3814–3827, 2013.

[BD14] S. Batabyal and S. S. Das. Analysis of a Call Admission Control Algorithm for Real-Time Traffic in OFDMA Based Cellular Networks. In *Proceedings of IEEE Vehicular Technology Conference VTC-Spring*, 1–5, 2014.

[BD15] S. Batabyal and S. S. Das. GoS Evaluation for OFDMA Cellular Networks. *Wireless Personal Communications*, 83(2):925–957, 2015.

[Bel01] M. G. Bellanger. Specification and design of a prototype filter for filter bank based multicarrier transmission. In *Proceedings of IEEE International Conference on Acoustics, Speech, and Signal Processing, 2001 (ICASSP '01)*, 4, 2417–2420, 2001.

[Bel10] M. Bellanger. Physical Layer for Future Broadband Radio Systems. In *2010 IEEE Radio and Wireless Symposium*, 436–439, 2010.

[BFE+04] B. Bosco, S. Franson, R. Emrick, S. Rockwell, and J. Holmes. A 60 GHz Transceiver Withmulti-Gigabit Data Rate Capability. In *Proceedings of IEEE Radio Wireless Conference*, 135–138, 2004.

[BH13] T. Bai and R. W. Heath Jr. Coverage Analysis for Millimeter Wave Cellular Networks with Blockage Effects. In *Proceedings of IEEE*

Global Conference on Signal and Information Processing (GlobalSIP), 2013.

[BLS⁺13] A. Benjebbour, A. Li, Y. Saito, Y. Kishiyama, Λ. Harada, and T. Nakamura. System-Level Performance of Downlink NOMA for Future LTE Enhancements. In *Globecom Workshops (GC Wkshps)*, IEEE, 66–70, 2013.

[Bos27] J. Bose. *Collected Physical Papers*. Longmans, Green and Co., 1927.

[Boy08] S. Boyd. *Convex Optimization II Lecture Notes*. Stanford University, 2008.

[BP03] T. Bonald and A. Proutière. Wireless Downlink Data Channels: User Performance and Cell Dimensioning. In *Proceedings of Annual International Conference on Mobile Computing and Networking, MobiCom*, 339–352, New York, USA, 2003, ACM.

[BPD16] S. Boddu, B. V. Philip, and S. S. Das. Analysis of Bandwidth Requirement of Users in Flexible Reuse Cellular Networks. *IEEE Communications Letters*, 20(3):614–617, 2016.

[BS14] J. Brady and A. Sayeed. Beamspace MU-MIMO for High-Density Gigabit Small Cell Access at Millimeter-Wave Frequencies. In *2014 IEEE 15th International Workshop on Signal Processing Advances in Wireless Communications (SPAWC)*, 80–84. IEEE, 2014.

[BS15] J. H. Brady and A. M. Sayeed. Differential Beamspace MIMO for High-Dimensional Multiuser Communication. In *2015 IEEE Global Conference on Signal and Information Processing (GlobalSIP)*, 310–314, IEEE, 2015.

[BSG+05] D. S. Baum, J. Salo, G. Del Galdo, M. Milojevic, P. Ky sti, and J. Hansen. An Interim Channel Model for Beyond-3G Systems. In *IEEE Vehicular Technology Conference*, Stockholm, Sweden, 2005.

[BV04] S. Boyd and L. Vandenberghe. *Convex Optimization*. Cambridge University Press, New York, NY, USA, 2004.

[C+03] C. C. Chong et al. A New Statistical Wideband Spatio-Temporal Channel Model for 5-GHz Band WLAN Systems. *IEEE Journal on Selected Areas in Communications*, 21(2):139–150, 2003.

[CA09] V. Chandrasekhar and J. Andrews. Spectrum Allocation in Tiered Cellular Networks. In *IEEE Transactions on Communications*, 57(10):3059–3068, 2009.

[CAG08] V. Chandrasekhar, J. Andrews, and A. Gatherer. Femtocell Networks: A Survey. In *IEEE Communications Magazine*, 46(9):59–67, 2008.

[CAG09] V. Chandrasekhar, J. G. Andrews, and A. Gatherer. Femtocell Networks: A Survey. *IEEE Communications Magazine*, 46(9):59–67, 2009.

[CCCG94] T. Cho, D. Cline, C. Conroy, and P. Gray. Design Considerations for Low-Power, High-Speed CMOS Analog/Digital Converters. In *Proceedings of IEEE Symposium Low Power Electronics*, 1994.

[CD14] P. Chandhar and S. S. Das. Area Spectral Efficiency of Co-channel Deployed OFDMA Femtocell Networks. In I*EEE Transactions on Wireless Communications*, 13(7):3524–3538, 2014.

[CD16] P. Chandhar and S. S. Das. Multi-Objective Framework for Dynamic Optimization of OFDMA Cellular Systems. In *IEEE Access*, 4:1889–1914, 2016.

[CEO02] G. Cherubini, E. Eleftheriou, and S. Olcer. Filtered Multitone Modulation for Very High-Speed Subscriber Lines. In *IEEE Journal on Selected Areas in Communications*, 20(5):1016–1028, 2002.

[CG01] S. T. Chung and A. J. Goldsmith. Degrees of Freedom in Adaptive Modulation: A Unified View. In *IEEE Transactions on Communications*, 49(9):1561–1571, 2001.

[CGKW97] P. Calegari, F. Guidec, P. Kuonen, and D. Wagner. Genetic Approach to Radio Network Optimization for Mobile Systems. In *IEEE 47th Vehicular Technology Conference*, 2, 755–759, 1997.

[Cha66] R. W. Chang. Synthesis of Band-Limited Orthogonal Signals for Multi-Channel Data Transmission. *Bell System Technical Journal*, 45(10):1775–1796, 1966.

[Cho14] J. Choi. Non-orthogonal Multiple Access in Downlink Coordinated Two-Point Systems. In *IEEE Communications Letters*, 18(2):313–316, 2014.

[CHS07] H. Claussen, L. T. W. Ho, and L. Samuel. Financial Analysis of a Pico-cellular Home Network Deployment. In *Proceedings of IEEE International Conference on Communications*, 5604–5609, 2007.

[Cis17] Cisco. *Cisco Visual Networking Index: Global Mobile Data Traffic Forecast Update, 2016–2021.* Technical Report, CISCO, 2017.

[CLCPRAHV14] S. L. Castellanos-Lopez, F. A. Cruz-Perez, M. E. Rivero-Angeles, and G. Hernandez-Valdez. Joint Connection Level and Packet Level Analysis of Cognitive Radio Networks with VoIP Traffic. In *IEEE Journal on Selected Areas in Communications*, 32(3):601–614, 2014.

[CRY+11] E. Cianca, T. Rossi, A. Yahalom, Y. Pinhasi, J. Farselotu, and C. Sacchi. EHF for Satellite Communications: The New Broadband Frontier. In *Proceedings of the IEEE*, 99(11):1858–1881, 2011.

[CSRL01] T. H. Cormen, C. Stein, R. L. Rivest, and C. E. Leiserson. *Introduction to Algorithms*. McGraw-Hill Higher Education, 2nd edition, 2001.

[CT06] T. M. Cover and J. A. Thomas. *Elements of Information Theory (Wiley Series in Telecommunications and Signal Processing)*. Wiley-Interscience, 2006.

[CZB$^+$10] L. M. Correia, D. Zeller, O. Blume, Y. Jading, G. Auer, L. der Perre, D. Ferling, and I. Godor. Challenges and Enabling Technologies for Energy Aware Mobile Radio Networks. In *IEEE Communications Magazine*, 48(11):1–12, 2010.

[DCP07] S. S. Das, E. D. Carvalho, and R. Prasad. Variable Sub-carrier Bandwidth in OFDM Framework. *Electronics Letters*, 43(1):46–47, 2007.

[DCP08] S. S. Das, E. D. Carvalho, and R. Prasad. Performance Analysis of OFDM Systems with Adaptive Sub Carrier Bandwidth. In *IEEE Transactions on Wireless Communications*, 7(4):1117–1122, 2008.

[DCR+15] M. De Sanctis, E. Cianca, T. Rossi, C. Sacchi, L. Mucchi, and R. Prasad. Waveform Design Solutions for EHF Broadband Satellite Communications. In *IEEE Communications Magazine*, 53(3):18–23, 2015.

[Deb01] K. Deb. *Multi-objective Optimization Using Evolutionary Algorithms*. Wiley, 2001.

[Dem02] A. Demir. Phase Noise and Timing Jitter in Oscillators with Colored-Noise Sources. In *IEEE Transactions on Circuits and Systems I: Fundamental Theory and Applications*, 49(12):1782–1791, 2002.

[DES+04] C. H. Doan, S. Emami, D. A. Sobel, A. M. Niknejad, and R. W. Brodersen. Design Considerations for 60 GHz CMOS Radios. In *IEEE Communications Magazine*, 42(12):132–140, 2004.

[DF14] R. Datta and G. Fettweis. Improved ACLR by Cancellation Carrier Insertion in GFDM Based Cognitive Radios. In *IEEE 79th Vehicular Technology Conference (VTC Spring),* 1–5, 2014.

[DG83] T. De Mazancourt and D. Gerlic. The inverse of a block-circulant matrix. In *IEEE Transaction on Antennas and Propagation*, 31, 5: 808–810, 1983.

[DGBA12a] H. S. Dhillon, R. K. Ganti, F. Baccelli, and J. G. Andrews. Modeling and analysis of k-tier downlink heterogeneous cellular networks. In *IEEE Journal on Selected Areas in Communications*, 30(3):550–560, 2012.

[DGBA12b] H. S. Dhillon, R. K. Ganti, F. Baccelli, and J. G. Andrews. Modeling and Analysis of k-tier Downlink Heterogeneous Cellular Networks. In *IEEE Journal on Selected Communications*, 30(3):550–560, 2012.

[DMLFda] R. Datta, N. Michailow, M. Lentmaier, and G. Fettweis. GFDM Interference Cancellation for Flexible Cognitive Radio PHY Design. In *IEEE Vehicular Technology Conference (VTC Fall)*, 1–5, 2012 (QC, Canada).

[DMRH10] R. Daniels, J. Murdock, T. S. Rappaport, and R. Heath. 60 GHz wireless: Up close and personal. *IEEE Microwave Magazine*, 11(7): 44–50, 2010.

[DNI+11] M.-T. Dao, V.-A. Nguyen, Y.-T. Im, S.-O. Park, and G. Yoon. 3D Polarized Channel Modeling and Performance Comparison of MIMO Antenna Configurations with Different Polarizations. In *IEEE Transactions on Antennas and Propagation*, 59(7):2672–2682, 2011.

[DR05] S. S. Das and I. Rahman. *Method for Transmitting Data in a Wireless Network*, Indian Patent, Granted Number 260999, 2005.

[DRW+09] K. Doppler, M. Rinne, C. Wijting, C. B. Ribeiro, and K. Hugl. Device-to-Device Communication as an Underlay to LTE-Advanced Networks. In *IEEE Communications Magazine*, 47(12): 42–49, 2009.

[DS14] H. Deng and A. Sayeed. Mm-Wave MIMO Channel Modeling and User Localization Using Sparse Beamspace Signatures. In *2014 IEEE 15th International Workshop on Signal Processing Advances in Wireless Communications (SPAWC)*, 130–134. IEEE, 2014.

[DSG+15] Z. Dawy, W. Saad, A. Ghosh, J. G. Andrews, and Elias Yaacoub. Towards Massive Machine Type Cellular Communications. *arXiv:1512.03452 [cs, math]*, 2015. arXiv: 1512.03452.

[DT15] S. S. Das and S. Tiwari. Discrete Fourier Transform Spreading-Based Generalised Frequency Division Multiplexing. *Electronics Letters*, 51(10):789–791, 2015.

[DZH15] N. Deng, W. Zhou, and M. Haenggi. The Ginibre Point Process as a Model for Wireless Networks with Repulsion. In *IEEE Transactions on Wireless Communications*, 14(1):107–121, 2015.

[E+04] V. Ecreg et al. *TGn Channel Models*. Technical Report, *IEEE Document*, 2004.

[EARAS+14] O. El Ayach, S. Rajagopal, S. Abu-Surra, Z. Pi, and R. W. Heath. Spatially Sparse Precoding in Millimeter Wave MIMO Systems.

In *IEEE Transactions on Wireless Communications*, 13(3):1499–1513, 2014.

[EDC14] M. Elkashlan, T. Q. Duong, and H. Chen. Millimeter Wave Communications for 5G: Fundamentals: Part I [Guest Editorial]. In *IEEE Communications Magazine*, 52–54, 2014.

[EDC15] M. Elkashlan, T. Q. Duong, and H. Chen. Millimeter Wave Communications for 5G- Part 2 : Applications [Guest Editorial]. In *IEEE Communications Magazine*, 53(1):166–167, 2015.

[ET08] ECMA standard 387 ECMC TC48. *High Rate 60 GHz PHY, MAC and HDMI PAL*. Technical Report, IEEE, 2008.

[Fal11] D. D. Falconer. Linear Precoding of OFDMA Signals to Minimize Their Instantaneous Power Variance. In *IEEE Transactions on Communication*, 59(4):1154–1162, 2011.

[FBGY10] B. Farhang-Boroujeny and C. Him (George) Yuen. Cosine Modulated and Offset QAM Filter Bank Multicarrier Techniques: A Continuous-Time Prospect. *EURASIP Journal on Advances in Signal Processing*, 2010:1–17, 2010.

[FBZ$^+$10] D. Ferling, T. Bohn, D. Zeller, P. Frenger, I. Godor, Y. Jading, and W. Tomaselli. Energy efficiency approaches for radio nodes. In *Future Network and Mobile Summit, 2010*, 1–9, 2010.

[FCC04] Federal Communications Commission DA 04-3151. 2004.

[FDM$^+$12] G. Fodor, E. Dahlman, G. Mildh, S. Parkvall, N. Reider, G. Miklos, and Z. Turanyi. Design Aspects of Network Assisted Device-to-Device Communications. In *IEEE Communications Magazine*, 50(3):170–177, 2012.

[Feh81] K. Feher. *Digital Communications - Satellite/Earth Station Engineering*. Prentice-Hall Inc., 1981.

[Fet14] G. P. Fettweis. The Tactile Internet: Applications and Challenges. In *IEEE Vehicular Technology Magazine*, 9(1):64–70, 2014.

[FG98] G. J. Foschini and M. J. Gans. On Limits of Wireless Communications in Fading Environment When Using Multiple Antennas. *Wireless Personal Communications*, 6:311–335, 1998.

[FJL$^+$13] D. Feng, C. Jiang, G. Lim, Jr. Cimini, L. J., Gang Feng, and G. Y. Li. A Survey of Energy-Efficient Wireless Communications. In *IEEE Communications Surveys and Tutorials*, 15(1):167–178, 2013.

[FKR13] R. Ford, C. Kim, and S. Rangan. Optimal User Association in Cellular Networks with Opportunistic Third-Party Backhaul. *arXiv preprint*, 2013.

[Fri46] H. T. Friis. A Note on a Simple Transmission Formula. In *Proceedings of the I.R.E. and Waves and Electrons*, 254–256, 1946.

[G+10] F. Gianesello et al. 325GHz CPW Band Pass Filter Integrated in Advanced HR SOI RF CMOS Technology. In *Proceedings of the 40th European Microwave Conference*, Paris, France, 2010.

[GAPR09] F. Gutierrez, S. Agarwal, K. Parrish, and T. S. Rappaport. On-Chip Integrated Antenna Structures in CMOS for 60 GHz WPAN systems. In *IEEE Journal on Selected Areas in Communications*, 27(8):1367–1378, 2009.

[Gar01] A. Garcia Armada. Understanding the Effects of Phase Noise in Orthogonal Frequency Division Multiplexing (OFDM). In *IEEE Transactions on Braodcasting*, 47(2):153–159, 2001.

[GDC+16] X. Gao, L. Dai, Z. Chen, Z. Wang, and Z. Zhang. Near-Optimal Beam Selection for Beamspace Mmwave Massive MIMO Systems. In *IEEE Communications Letters*, 20(5):1054–1057, 2016.

[GDM+15] Z. Gao, L. Dai, D. Mi, Z. Wang, M. A. Imran, and M. Z. Shakir. MmWave Massive MIMO Based Wireless Backhaul for 5G Ultra-Dense Network. *arXiv:1508.03940 [cs.IT]*, 2015.

[GH16] R. K. Ganti and M. Haenggi. Asymptotics and Approximation of the SIR Distribution in General Cellular Networks. In *IEEE Transactions on Wireless Communications*, 15(3):2130–2143, 2016.

[GI10] L. Giupponi and C. Ibars. Distributed Interference Control in OFDMA-Based Femtocells. In *2010 IEEE 21st International Symposium on Personal Indoor and Mobile Radio Communications (PIMRC)*, 1201–1206, 2010.

[GK13] H. Ghozlan and G. Kremer. On Wiener Phase Noise Channels at High Signal-to-Noise Ratio. In *IEEE International Seminar on Information Theory*, 2013.

[GKH+07] D. Gesbert, M. Kountouris, R. W. Heath Jr., C. Chae, and Thomas S lzer. From Single user to Multiuser Communications: Shifting the MIMO paradigm. In *IEEE Signal Processing Magazine*, 2007.

[GKZV09] S. Geng, J. Kivinen, X. Zhao, and P. Vainikainen. Millimeter-Wave Propagation Channel Characterization for Short Range Wireless Communication. In *IEEE Transactions on Vehicular Technology*, 58(1):3–13, 2009.

[GLR99] F. Gianetti, M. Luise, and R. Reggiannini. Mobile and Personal Communications in the 60 GHz Band: A Survey. *Wireless Personal Communications*, 10(2):207–243, 1999.

[GMP09] A. Golaup, M. Mustapha, and L. B. Patanapongpibul. Femtocell Access Control Strategy in UMTS and LTE. In *IEEE Communications Magazine*, 47(9):117–123, 2009.

[Gol05] A. J. Goldsmith. *Wireless Communications*. Cambridge University Press, 2005.

[GRA15] L. Gavrilovska, V. Rakovic, and V. Atanasovski. Visions Towards 5G: Technical Requirements and Potential Enablers. *Wireless Pers. Communications*, 1–27, 2015.

[GS91] A. Ghorbani and M. Sheikhan. The effect of Solid State Power Amplifiers (SSPAs) Nonlinearities on MPSK and M-QAM Signal Transmission. In *Sixth International Conference on Digital Processing of Signals in Communications*, 193–197, 1991.

[GSJZ09] F. Gordejuela-Sanchez, A. Juttner, and Jie Zhang. A Multiobjective Optimization Framework for IEEE 802.16e Network Design and Performance Analysis. In *IEEE Journal on Selected Areas in Communications*, 27(2):202–216, 2009.

[GSTH08] D. Gross, J. F. Shortle, J. M. Thompson, and C. M. Harris. *Fundamentals of Queueing Theory*, 4th edition. Wiley-Interscience, New York, NY, USA, 2008.

[Gus14] C. Gustafson. *60 GHz Wireless Propagation Channels: Characterization, Modeling and Evaluation*. Ph.D. thesis, Lund University, 2014.

[H+10] J. Hacker et al. THz MMICs based on InP HBT Technology. In *International Microwave Symposium*, Anaheim, California, USA, 2010.

[H+16a] R. He et al. Channel Measurements and Modeling for 5G Communication Systems at 3.5 GHz Band. In *URSI Asia-Pacific Radio Science Conference (URSI AP-RASC)*, Seoul, South Korea, 2016.

[H+16b] S. Hur et al. Proposal on Millimeter-Wave Channel Modeling for 5G Cellular System. In *IEEE Journal of Selected Topics in Signal Processing*, 10(3):454–469, 2016.

[Han15] K. Haneda. Channel Models and Beamforming at Millimeter-Wave Frequency Bands. In *IEICE Transactions on Communications*, 755–772, 2015.

[Har14] Y. Lan, A. Benjebbour, A. Li, A. Harada. Efficient and Dynamic Fractional Frequency Reuse for Downlink Non-Orthogonal Multiple Access. In *IEEE 79th Vehicular Technology Conference (VTC Spring)*, 1–5, 2014.

[Hat80] M. Hata. Empirical Formula for Propagation Loss in Land Mobile Radio Service. In *IEEE Transactions on Vehicular Technology*, 29(3):317–325, 1980.

[HB94] S. Hussain and S. K. Barton. The Performance of Coherent BPSK in the Presence of Oscillator Phase Noise for ka-Band Pico-Terminal Inbound Links. *Wireless Personal Communications*, 1(2):111–115, 1994.

[HBB11] Z. Hasan, H. Boostanimehr, and V. K. Bhargava. Green Cellular Networks: A Survey, Some Research Issues and Challenges. In *IEEE Communications Surveys and Tutorials*, 13(4):524–540, 2011.

[HBK+16] S. Hur, S. Baek, B. Kim, Y. Chang, A. F Molisch, T. S. Rappaport, K. Haneda, and J. Park. Proposal on Millimeter-Wave Channel Modeling for 5G Cellular System. In *IEEE Journal of Selected Topics in Signal Processing*, 10(3):454–469, 2016.

[HGPR+16] R. W. Heath Jr., N. Gonz lez-Prelcic, S. Rangan, W. Roh, and A. M. Sayeed. An Overview of Signal Processing Techniques for Millimeter Wave MIMO Systems. In *IEEE Journal of Selected Topics in Signal Processing*, 10(3):436–453, 2016.

[HH15] M. Hasan and E. Hossain. Distributed Resource Allocation in D2D-Enabled Multi-tier Cellular Networks: An Auction Approach. In *IEEE International Conference Communications,* 2949–2954, 2015.

[HH69] L. A. Hoffman and K. H. Huribut. A 94-GHz RADAR for Space Object Identification. In *G-MTT International Microwave Symposium*, 1969.

[Hir81] B. Hirosaki. An Orthogonally Multiplexed QAM System Using the Discrete Fourier Transform. In *IEEE Transactions on Communications*, 29(7):982–989, 1981.

[HK13] K. Higuchi and Y. Kishiyama. Non-Orthogonal Access with Random Beamforming and Intra-beam SIC for Cellular MIMO Downlink. In *IEEE 78th Vehicular Technology Conference (VTC Fall)*, 1–5, 2013.

[HKL+13] S. Hur, T. Kim, D. J. Love, J. V. Krogmeier, T. A. Thomas, and A. Ghosh. Millimeter Wave Beamforming for Wireless Backhaul and Access in Small Cell Networks. In *IEEE Transactions on Communications,* 61(10):4391–4403, 2013.

[HL16] T. Halsig and B. Lankl. Channel Parameter Estimation for LOS MIMO Systems. In *WSA*, 98–102, 2016.

[HT09] H. Holma and A. Toskala. *LTE for UMTS-OFDMA and SC-FDMA Based Radio Access*. John Wiley and Sons, 2009.

[HT10] H. Holma and A. Toskala. *LTE for UMTS: Evolution to LTE-Advanced*. John Wiley and Sons (Hoboken, NJ, USA), 2010.

[Hua10] Huawei. *Opportunities for Energy Savings in LTE Networks*. Technical Report, 3GPP TSG RAN WG1 Meeting #59bis, 3GPP-LTE, Valencia, Spain, 2010.

[Hui88] J. Y. Hui. Resource Allocation for Broadband Networks. In *IEEE Journal on Selected Areas in Communications*, 6(9):1598–1608, 1988.

[HWWS14] X. Hong, J. Wang, C.-X. Wang, and J. Shi. Cognitive radio in 5g: A Perspective on Energy-Spectral Efficiency Trade-off. In *IEEE Communications Magazine*, 52(7):46–53, 2014.

[HZJ+] Y. Han, H. Zhang, S. Jin, X. Li, R. Yu, and Y. Zhang. Investigation of Transmission Schemes for Millimeter-Wave Massive MU-MIMO Systems. In *IEEE Systems Journal*, 11, 72–83, 2017.

[iee09] Part 15.3: *Wireless Medium Access Control (MAC) and Physical Layer (PHY) Specifications for High Rate Wireless Personal Area Networks (WPANs), Amendment 2:Millimeter-wave-based Alternative Physical Layer Extension*. Technical Report, IEEE, 2009.

[IH09] T. Ingason and L. Haonan. *Line-of-Sight MIMO for Microwave Links*. Master's thesis, Chalmers University of Technology, Goeteborg, Sweden, 2009.

[IL09] A. Ikhlef and J. Louveaux. An enhanced mmse per subchannel equalizer for highly frequency selective channels for fbmc/oqam systems. In *IEEE 10th Workshop on Signal Processing Advances in Wireless Communications, 2009. SPAWC '09*, 186–190, 2009.

[IR08] ITU-R. *Guidelines for Evaluation of Radio Interface Technologies for IMT-Advanced*. Technical Report, M.2135, ITU-R, 2008.

[IRH+14] Chih-Lin I, C. Rowell, S. Han, Z. Xu, G. Li, and Z. Pan. Toward Green and Soft: A 5G Perspective. In *IEEE Communications Magazine*, 66–73, 2014.

[ISR05] T. Ihalainen, T. H. Stitz, and M. Renfors. Efficient Per-Carrier Channel Equalizer for Filter Bank Based Multicarrier Systems. In *IEEE International Symposium on Circuits and Systems, 2005. ISCAS 2005*, 3175–3178, 2005.

[ISRR07] T. Ihalainen, T. H. Stitz, M. Rinne, and M. Renfors. Channel Equalization in Filter Bank Based Multicarrier Modulation for Wireless Communications. *EURASIP Journal on Applied Signal Processing*, 2007(1):140–140, 2007.

[itu] Rec. ITU-R. *P.838-3 RECOMMENDATION ITU-R P.838-3 Specific attenuation Model for Rain for Use in Prediction Methods.* Technical Report, ITU, 2005.

[itu09] ITU-R M.2135-1. *Guidelines for Evaluation of Radio Interface Technologies for IMT-Advanced.* Technical Report, ITU, 2009.

[J+13] M. Jacob et al. Extension and Validation of the IEEE 802.11ad 60 GHz Human Blockage Model. In *Proceedings of European Conference on Antennas and Propagation*, 2806–2810, 2013.

[JB05] Y. Jin and J. Branke. Evolutionary Optimization in Uncertain Environments-A Survey. In *IEEE Transactions on Evolutionary Computation*, 9(3):303–317, 2005.

[JB68] E. C. Jordan and K. G. Balmain. *Electromagnetic Waves and Radiating Systems.* Prentice-Hall, 1968.

[JP03] S. K. Jayaweera and H. V. Poor. MIMO Capacity Results for Rician Fading Channels. In *Proceedings of Global Telecommunications Conference (GLOBECOM)*, 2003.

[JPP00] A Jalali, R. Padovani, and R. Pankaj. Data Throughput of CDMA-HDR a High Efficiency-High Data Rate Personal Communication Wireless System. In *IEEE 51st VTC*, 2000.

[JSXA12] H.-S. Jo, Y. J. Sang, P. Xia, and J. G. Andrews. Heterogeneous Cellular Networks with Flexible Cell Association: A Comprehensive Downlink SINR Analysis. In *IEEE Transactions on Wireless Communications*, 11(10):3484–3495, 2012.

[K+07] P. Ky sti et al. *WINNER II Channel Models.* Technical Report, European Commission, 2007.

[K+08] P. Kyosti et al. *WINNER II Channel Models D1.1.2.* Technical Report, European Union's 6th Framework Programme, 2008.

[KA06] M. Kang and M-Slim Alouini. Capacity of MIMO Rician Channels. In *IEEE Transactions in Communications*, 5(1):112–122, 2006.

[KBM+15] J. Kerttula, Y. Beyene, N. Malm, L. Zhou, K. Ruttik, O. Tirkkonen, and R. Jntti. Spectrum Sharing in D2D Enabled Hetnet. In *IEEE International Symposium on Dynamic Spectrum Access Networks (DySPAN)*, 2015, 267–268, 2015.

[KKA+14] A. Kammoun, H. Khanfir, Z. Altman, M. Debbah, and M. Kamoun. Preliminary Results on 3D Channel Modeling: From Theory to Standardization. In *IEEE Journal on Selected Areas in Communications*, 32(6):1219–1229, 2014.

[Kle75] L. Kleinrock. *Queueing Systems, volume I: Theory.* Wiley Interscience, 103–105, 1975.

[KLNK14] Y. Kawamoto, J. Liu, H. Nishiyama, and N. Kato. An Efficient Traffic Detouring Method by Using Device-to-Device Communication Technologies in Heterogeneous Network. In *IEEE Wireless Communications and Networking Conference (WCNC)*, 2162–2167, 2014.

[KM14] T. Kobayashi and N. Miyoshi. Uplink Cellular Network Models with Ginibre Deployed Base Stations. In *International Teletraffic Congress*, 1–7, 2014.

[Kob81] T. Han, K. Kobayashi. A New Achievable Rate Region for the Interference Channel. In *IEEE Transactions on Information Theory*, 27(1):49–60, 1981.

[KP14] M. Kountouris and N. Pappas. Approximating the Interference Distribution in Large Wireless Networks. In *International Symposium on Wireless Communications Systems,* 80–84, 2014.

[Kri15] R. Krishnan. *On the Impact of Phase Noise in Communication Systems - Performance Analysis and Algorithms*. Ph.D. thesis, Chalmers University of Technology, 2015.

[KS16] S. Kutty and D. Sen. Beamforming for Millimeter Wave Communications: An Inclusive Survey. In *IEEE Communications Surveys and Tutorials*, 18(2):949–973, Second Quarter 2016.

[KYF09] D. Knisely, T. Yoshizawa, and F. Favichia. Standardization of Femtocells in 3GPP. In *IEEE Communications Magazine*, 47(9):68–75, 2009.

[L+12] L. Liu et al. The COST 2100 MIMO Channel Model. In *IEEE Wireless Communications Magazine*, 92–99, 2012.

[LETM13] E. G. Larsson, O. Edfors, F. Tufvesson, and T. L. Marzetta. Massive MIMO for Next Generation Wireless Systems. In *IEEE Communications Magazine*, 52(2):186–195, 2013.

[LG01] L. Li and A. J. Goldsmith. Capacity and Optimal Resource Allocation for Fading Broadcast Channels .I. Ergodic Capacity. In *IEEE Transactions on Information Theory*, 47(3):1083–1102, 2001.

[LGS14] H. Lin, M. Gharba, and P. Siohan. Impact of Time and Carrier Frequency Offsets on the FBMC/OQAM Modulation Scheme. *Signal Processing*, 102:151–162, 2014.

[LHA06] L. B. Le, E. Hossain, and T. S. Alfa. Delay Statistics and Throughput Per-formance for Multi-rate Wireless Networks Under Multiuser Diversity. In *IEEE Transactions on Wireless Communications*, 5(11):3234–3243, 2006.

[LHK14] A. Li, A. Harada, and H. Kayama. A Novel Low Computational Complexity Power Assignment Method for Non-orthogonal Multiple

Access Systems. In *IEICE Transactions on Fundamentals of Electronics, Communications and Computer Sciences,* E97-A, 2014.

[LL02] K. Lieska and E. Laitinen. Optimization of GOS of Cellular Network. In *The 13th IEEE International Symposium on Personal, Indoor and Mobile Radio Communications*, 5, 2277–2281, 2002.

[LLL98] K. Lieska, E. Laitinen, and J. Lahteenmaki. Radio Coverage Optimization with Genetic Algorithms. In *The 9th IEEE International Symposium on Personal, Indoor and Mobile Radio Communications,* 1, 318–322 1, 1998.

[LSCP12] J. S. Lu, D. Steinbach, P. Cabrol, and P. Pietraski. Modeling Human Blockers in Millimeter Wave Radio Links. *ZTE Communications*, 10(4):23–28, 2012.

[LSE07] C. Liu, E. Skafidas, and R. J. Evans. Characterization of the 60 GHz Wireless Desktop Channel. In *IEEE Transactions on Antenna and Propagation*, 55(7):2129–2133, 2007.

[Lyn97] P. Lynch. The Dolph-Chebyshev Window: A Simple Optimal Filter. *Monthly weather review*, 125(4):655–660, 1997.

[LZG05] Q. Liu, S. Zhou, and G. B. Giannakis. Queuing with Adaptive Modulation and Coding Over Wireless Links: Cross-Layer Analysis and Design. In *IEEE Transactions on Wireless Communications*, 4(3):1142–1153, 2005.

[M+06] A. F. Molisch et al. A Comprehensive Standardized Model for Ultrawideband Propagation Channels. In *IEEE Journal on Antennas and Propagation*, 54(11):3151–3166, 2006.

[m2108] ITU. *Guidelines for Evaluation of Radio Interface Technologies for IMT-Advanced.* Technical Report, ITU, M2135, 2008.

[MAH+06] A. F. Molisch, H. Asplund, R. Heddergott, M. Steinbauer, and T. Zwick. The cost259 Directional Channel Model-Part I: Overview and Methodology. In *IEEE Transactions on Wireless Communications*, 5(12):3421–3433, 2006.

[Mal10] A. Maltsev. *Channel Models for 60 GHz WLAN Systems.* Technical Report, IEEE, 2010.

[MBCM09] M. A. Marsan, S. Buzzi, D. Ciullo, and M. Meo. Optimal Energy Savings in Cellular Access Networks, 1–5, 2009.

[MDCC17] S. Mukherjee, S. S. Das, A. Chatterjee, and S. Chatterjee. Analytical Calculation of Rician K-Factor for Indoor Wireless Channel Models. In *IEEE Access*, 5:19194–19212, 2017.

[MG09] H. A. Mahmoud and I. Guvenc. A Comparative Study of Different Deployment Modes for Femtocell Networks. In *IEEE 20th International*

Symposium on Personal, Indoor and Mobile Radio Communications, 13–16, 2009.

[MGK⁺12] N. Michailow, I. Gaspar, S. Krone, M. Lentmaier, and G. Fettweis. Generalized Frequency Division Multiplexing: Analysis of an Alternative Multi-Carrier Technique for Next Generation Cellular Systems. In *2012 International Symposium on Wireless Communication Systems*, 171–175, 2012.

[MH14] J. Mo and R. W. Heath Jr. High SNR Capacity of Millimeter Wave MIMO Systems with One-Bit Quantization. In *Information Theory and Applications Workshop (ITA)*, San Diego, CA, 2014.

[MK00] N. Mori and H. Kita. Genetic Algorithms for Adaptation to Dynamic Environments - A Survey. In *Annual Conference of the IEEE Industrial Electronics Society*, 4, 2947–2952, 2000.

[MKLFda] N. Michailow, S. Krone, M. Lentmaier, and G. Fettweis. Bit Error Rate Performance of Generalized Frequency Division Multiplexing. In *2012 IEEE Vehicular Technology Conference (VTC Fall)*, 1–5, 2012 (Quebec City, Canada).

[MLGnd] H. G. Myung, J. Lim, and D. Goodman. Peak-To-Average Power Ratio of Single Carrier FDMA Signals with Pulse Shaping. In *2006 IEEE International Symposium on Personal, Indoor and Mobile Radio Communications (PIMRC)*, 1–5, 2006 (Helsinki, Finland).

[MMG⁺14] N. Michailow, M. Matthe, I. S. Gaspar, A. N. Caldevilla, L. L. Mendes, A. Fes-tag, and G. Fettweis. Generalized Frequency Division Multiplexing for 5th Generation Cellular Networks. *IEEE Transactions on Communications*, 62(9):3045–3061, 2014.

[MML+11] A. Maltsev, R. Maslennikov, A. Lomayev, A. Sevastyananov, and A. Khoryaev. Statistical Channel Model for 60 GHz WLAN Systems in Conference Room Environment. *Radioengineering*, 20(2):409–422, 2011.

[MMS+09] A. Maltsev, R. Maslennikov, A. Sevastyananov, A. Khoryaev, and A. Lomayev. Experimental Investigations of 60 GHz WLAN Systems in Office Environment. *IEEE Journal on Selected Areas in Communication*, 27(8):1488–1499, 2009.

[MMSL10] A. Maltsev, R. Maslennikov, A. Sevastyanov, and A. Lomayev. Characteristics of Indoor Millimeter-Wave Channel at 60 GHz in Application to Perspective WLAN system. In *2010 Proceedings of the Fourth European Conference on Antennas and Propagation (EuCAP)*, Barcelona, Spain, 2010.

[MMW+15] H. Mehrpouyan, M. Matthaiou, R. Wang, G. K. Karagiannidis, and Y. Hua. Hybrid Millimeter-Wave Systems: A Novel Paradigm for HetNets. In *IEEE Communication Magazine*, 2015.

[MPM+10] A. Maltsev, E. Perahia, R. Maslennikov, A. Sevastyananov, A. Lomayev, and A. Khoryaev. Impact of Polarization Characteristics on 60-GHz Indoor Radio Communication Systems. In *IEEE Antenna and Propagation Letters*, 9:413–416, 2010.

[MR14] J. Murdock and T. S. Rappaport. Consumption Factor and Power-Efficiency Factor: A Theory for Evaluating the Energy Efficiency of Cascaded Communication Systems. In *IEEE Journal of Selected Areas in Communications*, 2014.

[MRSD15] G. R. MacCartney Jr., T. S. Rappaport, S. Sun, and S. Deng. Indoor Office Wideband Millimeter-Wave Propagation Measurements and Channel Models at 28 and 73 GHz for Ultra-Dense 5G Wireless Networks. In *IEEE Access*, 3:2388–2424, 2015.

[MS16] N. Miyoshi and T. Shirai. Spatial Modeling and Analysis of Cellular Networks Using the Ginibre Point Process: A Tutorial. In *IEICE Transactions*, 99-B(11):2247–2255, 2016.

[MSDP15] P. D. Mankar, B. R. Sahu, G. Das, and S. S. Pathak. Evaluation of Blocking Probability for Downlink in Poisson Networks. IN *IEEE Wireless Communications Letters*, 4(6):625–628, 2015.

[MSM+95] T. Manabe, K. Sato, H. Masuzawa, K. Taira, T. Ihara, Y. Kashashima, and K. Yamaki. Polarization Dependence of Multipath Propagation and High-speed Transmission Characteristics of Indoor Millimeter-Wave Channel at 60 GHz. In *IEEE Transactions on Vehicular Technology*, 44(2):268–274, 1995.

[MSS+14] H. A. Mustafa, M. Z. Shakir, Y. A. Sambo, K. A. Qaraqe, M. A. Imran, and E. Serpedin. Spectral Efficiency Improvements in Hetnets by Exploiting Device-to-Device Communications. In *Globecom Workshops*, 857–862, 2014.

[MTB12] D. Mattera, M. Tanda, and M. Bellanger. Frequency-Spreading Implementation of OFDM/OQAM Systems. In *2012 International Symposium on Wireless Communication Systems (ISWCS)*, 176–180, 2012.

[MTB13] D. Mattera, M. Tanda, and M. Bellanger. Performance Analysis of the Frequency-Despreading Structure for OFDM/OQAM Systems. In *IEEE 77th Vehicular Technology Conference (VTC Spring)*, 1–5, 2013.

[MTB15a] D. Mattera, M. Tanda, and M. Bellanger. Analysis of an FBMC/OQAM Scheme for Asynchronous Access in Wireless Communications. *EURASIP Journal on Advances in Signal Processing*, 2015(1):1–22, 2015.

[MTB15b] D. Mattera, M. Tanda, and M. Bellanger. Performance Analysis of Some Timing Offset Equalizers for FBMC/OQAM Systems. *Signal Processing*, 108:167–182, 2015.

[MTR00] H. Meunier, E.-G. Talbi, and P. Reininger. A Multiobjective Genetic Algorithm for Radio Network Optimization. In *Proceedings of the Congress on Evolutionary Computation*, 1, 317–324, 2000.

[MWI⁺09] C. Mehlführer, M. Wrulich, J. C. Ikuno, D. Bosanska, and M. Rupp. Simulating the Long Term Evolution Physical Layer. In *Proceedings of 17th European Signal Processing Conference*, 1471–1478, 2009.

[MWMZ07] N. B. Mehta, J. Wu, A. F. Molisch, and J. Zhang. Approximating a Sum of Random Variables with a Lognormal. In *IEEE Transactions on Wireless Communications*, 6(7):2690–2699, 2007.

[MZB01] H. Minn, M. Zeng, and V. K. Bhargava. On ARQ Scheme with Adaptive Error Control. In *IEEE Transactions on Vehicular Technology*, 50(6):1426–1436, 2001.

[MZSR17] Y. Medjahdi, R. Zayani, H. Shaek, and D. Roviras. WOLA Processing: A Useful Tool for Windowed Waveforms in 5g with Relaxed Synchronicity. In *2017 IEEE International Conference on Communications Workshops (ICC Workshops)*, 393–398, 2017.

[N+15] Vuokko Nurmela et al. *METIS Channel Models*. Technical Report, METIS Seventh Project Framework, 2015.

[NLJ+15] Y. Niu, Y. Li, D. Jin, L. Su, and A. V. Vasilakos. A Survey of Millimeter Wave (mmWave) Communications for 5G: Opportunities and Challenges. *Wireless Network*, 21(8):2657–2676, 2015.

[NMSR13] S. Nie, G. R. MacCartney Jr., S. Sun, and T. S. Rappaport. 72 GHz Millimeter Wave Indoor Measurements for Wireless and Backhaul Communications. In *IEEE 24th International Symposium on Personal, Indoor and Mobile Radio Communications: Mobile and Wireless Networks*, London, UK, 2013.

[NMSR14] S. Nie, G. R. MacCartney Jr., S. Sun, and T. S. Rappaport. 28 GHz and 73 GHz Signal Outage Study for Millimeter Wave Cellular and Backhaul Communications. In *Wireless Communications Symposium*, Washington DC, USA, 2014.

[NST+07] M. Narandzic, C. Schneider, R. Thom, T. J ms, P. Ky sti, and X. Zhao. Comparison of SCM, SCME and WINNER Channel Models. In *IEEE Vehicular Technology Conference*, 413–417, Dublin, Ireland, 2007.

[NSVP11] A. Nogueira, P. Salvador, R. Valadas, and A. Pacheco. *Markovian Modelling of Internet Traffic*. Springer Berlin Heidelberg, Berlin, Heidelberg, 98–124, 2011.

[NTOY96] T. Ninomiya, T. Saito, Y. Ohaebi, and H. Yatsuka. 60-GHz Transceiver for High-Speed Wireless LAN System. In *International Microwave Symposium Digest*, 1996.

[O+02] K. Ohata et al. Wireless 1.25 Gb/s Transceiver Module at 60 GHz-Band. In *IEEE International Solid-State Circuits Conference on Digest Technology Papers*, 1, 298–468, 2002.

[OBB$^+$14] A. Osseiran, F. Boccardi, V. Braun, K. Kusume, P. Marsch, M. Maternia, O. Queseth, M. Schellmann, H. Schotten, H. Taoka, H. Tullberg, M. A. Uusi-talo, B. Timus, and M. Fallgren. Scenarios for 5G Mobile and Wireless Communications: The Vision of the METIS Project. In *IEEE Communications Magazine*, 52(5):26–35, 2014.

[OKH12] N. Otao, Y. Kishiyama, and K. Higuchi. Performance of Non-orthogonal Access with SIC in Cellular Downlink Using Proportional Fair-Based Resource Allocation. In *2012 International Symposium on Wireless Communication Systems (ISWCS)*, 476–480, 2012.

[OOF68] T. Okumura, E. Ohmori, and K. Fukuda. Field Strength and its Variability in VHF and UHF Land Mobile Service. *Review Electrical Communication Laboratory*, 16(9–10):825–873, 1968.

[Ort08] S. Ortiz. The Wireless Industry Begins to Embrace Femtocells. In *IEEE Computer*, 41(7):14–17, 2008.

[P.12] P. Ghosh, and S. S. Das, and P. Chandhar. Estimation of Effective Radio Re-source Usage for VoIP Scheduling in OFDMA Cellular Networks. In *Proceedings of IEEE Vehicular Technology Conference VTC-Spring*, 1–6, 2012.

[Pal18] B. Palit. *Cross Layer Design of Schedulers for Quality of Service Provisioning of Multi-Class Traffic in Wireless Networks*. Ph.D. thesis, Indian Institute of Technology, Kharagpur, 2018

[PBL+13] F. Pepe, A. Bonfati, S. Levantino, C. Samori, and A. L. Lacaita. Suppression of Flicker Noise Up-Conversion in a 65-nm CMOS VCO in the 3.0-to-3.6 GHz Band. In *IEEE Transactions on Solid State Circuits*, 48(10):2375–2389, 2013.

[PBR+12] P. Pietraski, D. Britz, A. Roy, R. Pragada, and G. Charlton. Millimeter Wave and Terahertz Communications: Feasibility and Challenges. *ZTE Communications*, 10(4):3–12, 2012.

[PD14] P. Parida and S. S. Das. Power Allocation in OFDM Based NOMA Systems: A DC Programming Approach. In *2014 IEEE Globecom Workshops (GC Wkshps)*, 1026–1031, 2014.

[PD15] B. Palit and S. S. Das. Performance Evaluation of Mixed Traffic Schedulers in OFDMA Networks. *Wireless Pers. Communications*, 83(2):895–924, 2015.

[PKH+08] J. Puttonen, N. Kolehmainen, T. Henttonen, Martti Moisio, and M. Rinne. Mixed Traffic Packet Scheduling in UTRAN Long Term Evolution Downlink. In *Proceedings of IEEE International Symposium on Personal, Indoor and Mobile Radio Communications, PIMRC*, 1–5, 2008.

[PM06] J. G. Proakis and D. G. Manolakis. *Digital Signal Processing*. Pearson, 2006.

[PNG03a] A. Paulraj, R. Nabar, and D. Gore. *Introduction to Space-Time Wireless Communications*. Cambridge University Press, 2003.

[PNG03b] A. Paulraj, R. Nabar, and D. Gore. *Introduction to Space-Time Wireless Communications*. Cambridge University Press, 2003.

[PNG06] A. Paulraj, R. Nabar, and D. Gore. *Introduction to Space-Time Wireless Communications*. Cambridge University Press, 2006.

[PO08] T. Paul and T. Ogunfrunmiri. Wireless Lan Comes of Age: Understanding the IEEE 802.11n Amendment. In *IEEE Circuits and Systems Magazine*, 8(1):28–54, 2008.

[Poz05] D. M. Pozar. *Microwave Engineering*. Wiley, 2005.

[PP02] A. Papoulis and S. U. Pillai. *Probability, Random Variables, and Stochastic Processes*. Tata McGraw-Hill Education, 2002.

[PRB10] M. Poggioni, L. Rugini, and P. Banelli. QoS Analysis of a Scheduling Policy for Heterogeneous Users Employing AMC Jointly with ARQ. In *IEEE Transactions on Communications*, 58(9):2639–2652, 2010.

[PS08] J. G. Proakis and M. Salehi. *Digital Communication, Chapter 14*. McGraw Hill (New York, USA), 2008.

[PSML06] P. Pagani, I. Siaud, N. Malhouroux, and W. Li. *Adaptation of the France Telecom 60 GHz channel model to the TG3c Framework*. Technical Report, IEEE, 2006.

[PT12] S. Payami and F. Tufvesson. Channel Measurements and Analysis for Very Large Array Systems at 2.6 Ghz. In *2012 6th European Conference on Antennas and Propagation (EUCAP)*, 433–437, IEEE, 2012.

[PvL93] R. Prasad and B. C. van Lieshout. Cochannel Interference Probability for Micro and Picocellular Systems at 60 GHz. In *IEE Electronic Letters*, 29(22):1909–1910, 1993.

[PZ02] D. Parker and D. C. Zimmermann. Phased Arrays-Part I: Theory and Architectures. In *IEEE Transactions on Microwave Theory and Techniques*, 50(3):678–687, 2002.

[QL06] Z. Qingling and J. Li. Rain Attenuation in Millimeter Wave Ranges. In *7th International Symposium on Antennas, Propagation and EM Theory*, Guilin, China, 2006.

[Rap01] T. Rappaport. *Wireless Communications: Principles and Practice, 2nd edition*. Prentice Hall PTR, Upper Saddle River, NJ, USA, 2001.

[Rap02] T. S. Rappaport. *Wireless Communications*. Prentice Hall, 2002.

[Rap91] C. Rapp. Effects of HPA-Nonlinearity on a 4-DPSK/OFDM-Signal for a Digital Sound Broadcasting System. In *Proceedings of the Second European Conference on Satellite Communications*, 179–184, Liege, Belgium, October 22-24, 1991.

[RFF09] F. Richter, A. J. Fehske, and G. P. Fettweis. Energy Efficiency Aspects of Base Station Deployment Strategies for Cellular Networks. In *IEEE 70th Vehicular Technology Conference Fall (VTC 2009-Fall)*, 1–5, 2009.

[rGPPG17] 3rd Generation Partnership Project (3GPP). Technical Specification Group Radio Access Network; Evolved Universal Terrestrial Radio Access (E-UTRA); Further Advancements for E-UTRA Physical Layer Aspects (Release 9). TR 36.814, 2017.

[rGPPG18] 3rd Generation Partnership Project (3GPP). 3rd Generation Partnership Project; Technical Specification Group Radio Access Network; Study on 3d Channel Model for LTE (release 12). TR 36.873, 2018.

[RHHB09] M. M. Rashid, M. J. Hossain, E. Hossain, and V. K. Bhargava. Opportunistic Spectrum Scheduling for Multiuser Cognitive Radio: A Queueing Analysis. In *IEEE Transactions on Wireless Communications*, 8(10):5259–5269, 2009.

[RK09] J. S. Rieh and D. H. Kim. An Overview of Semiconductor Technologies and Circuits for Terahertz Communication Applications. In *Proceedings of IEEE GLOBECOM Workshops*, 1–6, Honolulu, Hawaii, USA, 2009.

[RLM02] J. Razavilar, K. J. R. Liu, and S. I. Marcus. Jointly Optimized Bitrate/Delay Control Policy for Wireless Packet Networks with Fading Channels. In *IEEE Transactions on Communications*, 50(3):484–494, 2002.

[RMG11] T. S. Rappaport, J. N. Murdock, and F. Gutierrez. State of the Art in 60-GHz Integrated Circuits and Systems for Wireless Communications. In *IEEE Communications Magazine*, 99(8):1390–1436, 2011.

[RPL+13] F. Rusek, D. Persson, B. K. Lau, E. G. Larsson, T. L. Marzetta, O. Edfors, and F.R. Tufvesson. Scaling up MIMO: Opportunities and challenges with very large arrays. In *IEEE Signal Process. Mag*, 40–60, 2013.

[RRE14] S. Rungan, T. S. Rappaport, and E. Erkip. Millimeter Wave Cellular Wireless Networks: Potentials and Challenges. In *Proceedings of the IEEE*, 102(3):366–385, 2014.

[RSM+13a] T. S. Rappaport, S. Sun, R. Mayzus, H. Zhao, Y. Azar, K. Wang, G. N. Wong, J. K. Schluz, M. Samimi, and F. Gutierrez. Millimeter Wave Mobile Communications for 5G Cellular: It Will Work! In *IEEE Access*, 1:335–349, 2013.

[RSM+13b] T. S. Rappaport, S. Sun, R. Mayzus, H. Zhao, Y. Azar, K. Wang, G. N. Wong, J. K. Schulz, M. Samimi, and F. Gutierrez. Millimeter Wave Mobile Communications for 5G Cellular: It Will Work! *IEEE Access*, 1:335–349, 2013.

[Sad06] A. Sadri. *Summary of the Usage Models for 802.15.3c*. Technical Report, IEEE 802.15-06-0369-09-003c, 2006.

[Sal67a] B. Saltzberg. Performance of an Efficient Parallel Data Transmission System. In *IEEE Transactions on Communication Technology*, 15(6):805–811, 1967.

[Sal67b] B. Saltzberg. Performance of an Efficient Parallel Data Transmission System. In *IEEE Transactions on Communication Technology*, 15(6):805–811, 1967.

[Sal81] A. A. M. Saleh. Frequency-Independent and Frequency-Dependent Nonlinear Models of TWT Amplifiers. *IEEE Transactions on Communications*, 29:1715–1720, 1981.

[Saw06] H. Sawada. *LOS Office Channels Based on the TSV Model*. Technical Report, IEEE, 2006.

[Say02] A. M. Sayeed. Deconstructing Multiantenna Fading Channels. In *IEEE Transactions on Signal Processing*, 50(10):2563–2579, 2002.

[SB15] A. M. Sayeed and J. H. Brady. High Frequency Differential MIMO: Basic Theory and Transceiver Architectures. In *2015 IEEE International Conference on Communications (ICC)*, 4333–4338. IEEE, 2015.

[SBKN13] Y. Saito, A. Benjebbour, Y. Kishiyama, and T. Nakamura. System-Level Performance Evaluation of Downlink Non-orthogonal Multiple Access (NOMA). In *IEEE 24th International Symposium on Personal Indoor and Mobile Radio Communications (PIMRC)*, 611–615, 2013.

[SCM+13] C. Stallo, E. Cianca, S. Mukherjee, T. Rossi, M. De Sanctis, and M. Ruggieri. UWB for Multi-Gigabit/s Communications Beyond 60 GHz. *Telecommunications Systems Journal*, 52(1):161–181, 2013.

[SDS15] T. Seyama, T. Dateki, and H. Seki. Efficient Selection of User Sets for Downlink Non-orthogonal Multiple Access. In *IEEE 26th Annual International Symposium on Personal, Indoor, and Mobile Radio Communications (PIMRC)*, 1062–1066, 2015.

[Shaq09] M. Shaqfeh, N. Goertz, and J. Thompson. "Ergodic capacity of block-fading Gaussian broadcast and multi-access channels for single-user-selection and constant-power," *2009 17th European Signal Processing Conference*, Glasgow, 2009, pp. 784–788.

[SHSB13] R. Sassioui, A. El Hamss, L. Szczecinski, and M. Benjillali. Resource Allocation for Downlink Channel Transmission Based on Superposition Coding. *CoRR*, abs/1305.0619, 2013.

[Shu04] D. Shutin. Cluster Analysis of Wireless Channel Impulse Responses. In *Zurich Seminar on Communications*, Zurich, Switzerland, 2004.

[SJJS00] Q. Spencer, B. D. Jeffs, M. A. Jensen, and A. L. Swindlehurst. Modeling of Statistical Time and Angle of Arrival Characteristics. In *IEEE Journal on Selected Areas in Communications*, 18(3):347–360, 2000.

[SJK+03] T. K. Sarkar, Z. Ji, K. Kim, A. Medouri, and M. Salazar-Palma. A Survey of Various Propagation Models for Mobile Communication. In *IEEE Antennas and Propagation Magazine*, 45(3):51–82, 2003.

[SKYK11] K. Son, H. Kim, Y. Yi, and B. Krishnamachari. Base Station Operation and User Association Mechanisms for Energy-Delay Tradeoffs in Green Cellular Networks. 29(8):1525–1536, 2011.

[Sli02] S. B. Slimane. Peak-to-Average Power Ratio Reduction of OFDM Signals Using Broadband Pulse Shaping. In *2002 IEEE Vehicular Technology Conference VTC Fall*, 2, 889–893, 2002.

[SM17] A. Chatterjee, S. Chatterjee, S. Mukherjee, S. S. Das. Analytical Calculation of Rician k-Factor for Indoor Wireless Channel Models. In *IEEE Access*, 5:19194–19212, 2017.

[SMB01] M. Steinbauer, A. F. Molisch, and E. Bonek. The Double-Directional Radio Channel. In *IEEE Transactions on Antennas and Propagation*, 43(4):51–63, 2001.

[SMC+12] C. Stallo, S. Mukherjee, E. Cianca, T. Rossi, M. De Sanctis, and M. Ruggieri. System Level Comparison of Broadband Satellite Communications in Ka/Q/W bands. In *IEEE First AESS European Conference on Satellite Communications*, Rome, Italy, 2012.

[Smu09] P. F. M. Smulders. Statistical Characterization of 60-GHz Indoor Radio Channels. In *IEEE Transactions on Antenna and Propagation*, 57(10):2820–2829, 2009.

[SNS+14] A. I. Sulyman, A. T. Nassar, M. K. Samimi, G. R. Maccartney, T. S. Rappaport, and A. Alsanie. Radio Propagation Path Loss Models for 5G Cellular Networks in the 28 Ghz and 38 Ghz Millimeter-Wave Bands. In *IEEE Communications Magazine*, 52(9):78–86, 2014.

[Spi77] J. J. Spilker. *Digital Communications by Satellite*. Prentice-Hall Inc., 1977.

[SR05] L. Schumacher and B. Raghothaman. Closed-Form Expressions for the Correlation Coefficient of Directive Antennas Impinged by a Multimodal Truncated Laplacian PAS. In *IEEE Transactions on Wireless Communications*, 4(4):1351–1359, 2005.

[SR16] M. K. Samimi and T. S. Rappaport. 3-D Millimeter-Wave Statistical Channel Model for 5G Wireless System Design. In *IEEE Transactions on Microwave Theory and Techniques*, 64(7):2207–2225, 2016.

[ST95] S. D. Sandberg and M. A. Tzannes. Overlapped Discrete Multitone Modulation for High Speed Copper Wire Communications. In *IEEE Journal on Selected Areas in Communications*, 13(9):1571–1585, 1995.

[sta] Available at: https://www.statista.com/.

[STB09] S. Sesia, I. Toufik, and M. Baker. *LTE – The UMTS Long Term Evolution: From Theory to Practice*. John Wiley and Sons, 2009.

[STB11] S. Sesia, I. Toufik, and M. Baker. *LTE - The UMTS Long Term Evolution: From Theory to Practice*. John Wiley and Sons (Hoboken, NJ, USA), 2011.

[Stu01] G. L. Stuber. *Principles of Mobile Communication (2nd Ed.)*. Kluwer Academic Publishers, Norwell, MA, USA, 2001.

[Stu11] G. Stueber. *Principles of Mobile Communication*. Springer-Verlag New York, 2011.

[SV87] A. A. M. Saleh and R. A. Valenzuela. A Statistical Model for Indoor Multipath Propagation. In *IEEE Transactions on Selected Areas in Communication*, SAC-5(2):128–137, 1987.

[SVTR09] V. Syrjala, M. Valkama, N. N. Tchamov, and J. Rinne. Phase Noise Modelling and Mitigation Techniques in OFDM Communications Systems. In *Proceedings of the 2009 conference on Wireless Telecommunications Symposium*, 2009.

[SW14] F. Schaich and T. Wild. Waveform Contenders for 5G 2014; OFDM vs. FBMC vs. UFMC. In *6th International Symposium on*

Communications, Control and Signal Processing (ISCCSP), 457–460, 2014.

[SWSQ15] H. Sun, M. Wildemeersch, M. Sheng, and T. Q. S. Quek. D2d Enhanced Heterogeneous Cellular Networks with Dynamic TDD. In *IEEE Transactions on Wireless Communications*, 14(8):4204–4218, 2015.

[SY12] I. Siomina and Di Yuan. Analysis of Cell Load Coupling for LTE Network Planning and Optimization. In *IEEE Transactions on Wireless Communications*, 11(6):2287–2297, 2012.

[SZM+09] S. Singh, F. Ziliotto, U. Madhow, E. M. Belding, and M. Rodwell. Blockage and Directivity in 60 GHz Wireless Personal Area Networks: From Cross-Layer Model to Multi Hop MAC Design. In *IEEE Journal on Selected Areas in Communications*, 27(8):1400–1413, 2009.

[TA14] N. Tadayon and S. Aissa. Multi-Channel Cognitive Radio Networks: Modeling, Analysis and Synthesis. In *IEEE Journal on Selected Areas in Communications*, 32(11):2065–2074, 2014.

[TDB15] S. Tiwari, S. S. Das, and K. K. Bandyopadhyay. Precoded Generalised Frequency Division Multiplexing System to Combat Inter-Carrier Interference: Performance Analysis. In *IET Communications*, 2015.

[tel] Available at: http://www.telecomlead.com/. [thi] Available at: https://www.thinksmallcell.com/.

[Tel99] I. E. Telatar. Capacity of Multi-antenna Gaussian Channels. *European Transactions of Communications*, 10:585–595, 1999.

[TG07] J. A. Tropp and A. C. Gilbert. Signal Recovery from Random Measurements via Orthogonal Matching Pursuit. In *IEEE Transactions on Information Theory*, 53(12):4655–4666, 2007.

[tg3] IEEE 802.15. *IEEE 802.15 WPAN Millimeter Wave Alternative PHY Task Group 3c (TG3c)*. Technical Report, IEEE.

[TNMR14a] T. A. Thomas, H. C. Nguyen, G. R. MacCartney, and T. S. Rappaport. 3D Mmwave Channel Model Proposal. In *IEEE 80th Vehicular Technology Conference (VTC Fall)*, 1–6, IEEE, 2014.

[TNMR14b] T. A. Thomas, H. C. Nguyen, G. R. MacCartney Jr., and T. S. Rappaport. 3D mmWave Channel Model Proposal. In *Vehicular Technology Conference*, 2014.

[TR~14] v1.3.0 TR 36.873. *Study on 3D Channel Model for LTE (Release 12)*. Technical Report, 3GPP, 2014.

[tr208] TR.25.820. *3G Home NodeB Study Item Technical Report*. Technical Report TR.25.820 v8.2.0, 3GPP, 2008.

[TR317] TR38.900. *Study on Channel Model for Frequency Spectrum Above 6 Ghz (Release 14):3rd Generation Partnership Project; Technical Specification Group Radio Access Network*. Technical Report, 3GPP, 2017.

[Tra73] G. E. Trapp. Inverses of Circulant Matrices and Block Circulant Matrices. *Kyungpook Math. J.*, 13(1):11–20, 1973.

[TS15] R. Taori and A. Sridharan. Point-to-Multipoint in-Band Mmwave Back-Haul for 5G Networks. In *IEEE Wireless Communications*, 53(1):195–201, 2015.

[Tse97] D. N. Tse. Optimal Power Allocation Over Parallel Gaussian Broadcast Channels. In *Proceedings of IEEE International Symposium on Information Theory*, 1997.

[Tse99] D. N. Tse. Optimal Power Allocation Over Parallel Gaussian Broadcast Channels. In *Proceedings of IEEE International Symposium on Information Theory*, 1999.

[Tui02] P. W. Tuinenga. Models Rush In Where Simulations Fear To Tread: Extending the Baseband-Equivalent Method. In *International Behavioral Modeling and Simulation Conference*, Santa Rosa, 2002.

[VDZF15] N. S. Vo, T. Q. Duong, H. J. Zepernick, and M. Fiedler. A Cross-Layer Optimized Scheme and Its Application in Mobile Multimedia Networks With QoS Provision. In *IEEE Systems Journal*, 99:1–14, 2015.

[VF15] C. Vlachos and V. Friderikos. Optimal Device-to-Device Cell Association and Load Balancing. In *IEEE International Conference Communications*, 5441–5447, 2015.

[VIS+09] A. Viholainen, T. Ihalainen, T. H. Stitz, M. Renfors, and M. Bellanger. Prototype Filter Design for Filter Bank Based Multicarrier Transmission. In *17th European Signal Processing Conference*, 1359–1363, 2009.

[VSS10] N. Vucic, S. Shi, and M. Schubert. DC Programming Approach for Resource Allocation in Wireless Networks. In *8th International Symposium on Modeling and Optimization in Mobile, Ad Hoc and Wireless Networks (WiOpt)*, 380–386, 2010.

[VWS+13a] V. Vakilian, T. Wild, F. Schaich, S. ten Brink, and J.-F. Frigon. Universal-Filtered Multi-Carrier technique for wireless systems beyond LTE. In *2013 IEEE Globecom Workshops*, 223–228, 2013.

[VWS+13b] V. Vakilian, T. Wild, F. Schaich, S. ten Brink, and J.-F. Frigon. Universal-Filtered Multi-carrier Technique for Wireless Systems Beyond LTE. In *Globecom Workshops (GC Wkshps), 2013 IEEE*, 223–228, 2013.

[WB88] J. Walfisch and H. L. Bertoni. A Theoretical Model of UHF Propagation in Urban Environments. In *IEEE Transactions on Antenna and Propagation*, 36(12):1788–1796, 1988.

[WBN08] D. S. Waldhauser, L. G. Baltar, and J. Nossek. MMSE Subcarrier Equalization for Filter Bank Based Multicarrier Systems. In *IEEE 9th Workshop on Signal Processing Advances in Wireless Communications, SPAWC 2008*, 525–529, 2008.

[WDZH16] H. Wei, N. Deng, W. Zhou, and M. Haenggi. Approximate SIR Analysis in General Heterogeneous Cellular Networks. In *IEEE Transactions on Communications*, 64(3):1259–1273, 2016.

[WHCW13] J. Wang, A. Huang, L. Cai, and W. Wang. On the Queue Dynamics of Multiuser Multichannel Cognitive Radio Networks. In *IEEE Transactions on Vehicular Technology*, 62(3):1314–1328, 2013.

[WHKJ04] T. Weiss, J. Hillenbrand, A. Krohn, and F. K. Jondral. Mutual Interference in OFDM-Based Spectrum Pooling Systems. In *2004 IEEE 59th Vehicular Technology Conference. VTC 2004-Spring (IEEE Cat. No.04CH37514)*, 4, 1873–1877, 2004.

[WKtB+13] G. Wunder, M. Kasparick, S. ten Brink, F. Schaich, T. Wild, I. Gaspar, E. Ohlmer, S. Krone, N. Michailow, A. Navarro, G. Fettweis, D. Ktenas, V. Berg, M. Dryjanski, S. Pietrzyk, and B. Eged. 5GNOW: Challenging the LTE Design Paradigms of Orthogonality and Synchronicity. In *2013 IEEE VTC Spring*, 1–5, 2013.

[WLSV15] P. Wang, Y. Li, L. Songa, and B. Vucetic. Multi-gigabit Millimeter Wave Wireless Communications for 5G: From Fixed Access to Cellular Networks. In *IEEE Communications Magazine*, 53(1):168–178, 2015.

[WM95] H. S. Wang and N. Moayeri. Finite-state Markov Channel-A Useful Model for Radio Communication Channels. In *IEEE Transactions on Vehicular Technology*, 44(1):163–171, 1995.

[WMMZ11] J. Wu, N. B. Mehta, A. F. Molisch, and J. Zhang. Unified Spectral Efficiency Analysis of Cellular Systems with Channel-Aware Schedulers. In *IEEE Transactions on Communications*, 59(12):3463–3474, 2011.

[WWA+14] S. Wu, C.-X. Wang, M. M. Alwakeel, Y. He, et al. A Non-stationary 3-d Wideband Twin-Cluster Model for 5G Massive MIMO

Channels. In *IEEE Journal on Selected Areas in Communications*, 32(6):1207–1218, 2014.

[WWH+15] S. Wu, C.-X. Wang, H. Haas, M. M. Alwakeel, B. Ai, et al. A Non-Stationary Wideband Channel Model for Massive MIMO Communication Systems. In *IEEE Transactions on Wireless Communications*, 14(3):1434–1446, 2015.

[WWS15] X. Wang, T. Wild, and F. Schaich. Filter Optimization for Carrier-Frequency- and Timing-Offset in Universal Filtered Multi-carrier Systems. In *IEEE 81st Vehicular Technology Conference (VTC Spring)*, 1–6, 2015.

[WWSFdS14] X. Wang, T. Wild, F. Schaich, and A. F. dos Santos. Universal Filtered Multi-carrier with Leakage-Based Filter Optimization. In *Proceedings of European Wireless 2014; 20th European Wireless Conference,* 1–5, 2014.

[WWStB15] X. Wang, T. Wild, F. Schaich, and S. ten Brink. Pilot-aided Channel Estimation for Universal Filtered Multi-carrier. In *IEEE 82nd Vehicular Technology Conference (VTC Fall),* 1–5, 2015.

[www10] Available at: www.femtoforum.org. Interference Management in OFDMA Femtocells. Available at: http://smallcellforum.org/smallcell forum/resources-white-papers#OFDMA, 2010.

[WZZW15] H. Wang, Z. Zhang, Y. Zhang, and C. Wang. Universal Filtered Multi-carrier Transmission with Active Interference Cancellation. In *International Conference on Wireless Communications Signal Processing (WCSP),* 1–6, 2015.

[XKR02] H. Xu, V. Kukshya, and T. S. Rappaport. Spatial and Temporal Characteristics of 60-GHz Indoor Channels. In *IEEE Journal in Selected Areas in Communications*, 20(3):620–630, 2002.

[Y. 13] Y. Saito et al. Non-orthogonal Multiple Access (NOMA) for Cellular Future Radio Access. In *IEEE 77th VTC Spring*, 1–5, 2013.

[Y+11] Y. Yu et al. *Integrated 60GHz RF Beamforming in CMOS*. Springer, 2011.

[YMPM08] Y. Fan, M. Kuusela, P. Lunden, and M. Valkama. Downlink VoIP Support for Evolved UTRA. In *Proceedings of IEEE Wireless Communications Network Conference,* 1933–1938, 2008.

[Yon07] S. K. Yong. *TG3c Channel Modeling Sub-committee Final Report*. Technical Report, IEEE, 2007.

[Yos] S. Yost. Mmwave: Battle of the Bands. *National Instruments White Paper,* 2016.

[YSH07] H. Yang, P. F. M. Smulders, and M. H. A. J. Herben. Channel Characterization and Transmission Performance for Various Channels at 60 GHz. *Eurasip Journal on Wireless Communication and Networking*, 2007.

[YTLK08] S.-P. Yeh, S. Talwar, S.-C. Lee, and H. Kim. WiMAX femtocells: A Perspective on Network Architecture, Capacity, and Coverage. In *IEEE Communications Magazine*, 46(10):58–65, 2008.

[YXVG11] S. Yong, P. Xia, and A. Valdes-Garcia. *60 GHz Technology For Gbps WLAN and WPAN*. Wiley, 2011.

[Z+13] H. Zhao et al. 28 GHz Millimeter Wave Cellular Communication Measurements for Reflection and Penetration Loss in and around Buildings in New York City. In *IEEE International Conference on Communications*, 5163–5167, Budapest, Hungary, 2013.

[ZIX+17] L. Zhang, A. Ijaz, P. Xiao, M. Molu, and R. Tafazolli. Filtered OFDM Systems, Algorithms and Performance Analysis for 5G and Beyond. In *IEEE Transactions on Communications*, 99:1–1, 2017.

[ZMSR16] R. Zayani, Y. Medjahdi, H. Shaiek, and D. Roviras. WOLA-OFDM: A Potential Candidate for Asynchronous 5G. In *2016 IEEE Globecom Workshops (GC Wkshps)*, 1–5, 2016.

[Zou09] H. Zou, A. Chowdhery and J. M. Cioffi. "A centralized multi-level water-filling algorithm for Dynamic Spectrum Management," *2009 Conference Record of the Forty-Third Asilomar Conference on Signals, Systems and Computers*, Pacific Grove, CA, 2009, pp. 1101–1105. doi: 10.1109/ACSSC.2009.5470058

[ZOY14] K. Zheng, S. Ou, and X. Yin. Massive MIMO Channel Models: A Survey. *International Journal of Antennas and Propagation*, 1–10, 2014.

[ZSAA13] A. Zafar, M. Shaqfeh, M.-S. Alouini, and H. Alnuweiri. On Multiple Users Scheduling Using Superposition Coding Over Rayleigh Fading Channels. In *IEEE Communications Letters*, 17(4):733–736, 2013.

[ZVM10] H. Zhang, S. Venkateswaran, and U. Madhow. Channel Modeling and MIMO Capacity for Outdoor Millimeter Wave Links. In *Proceedings of IEEE Wireless Communications and Networking Conference (WCNC)*, 2010.

Index

About the Authors

Suvra Sekhar Das, completed his Ph.D. from Aalborg University, Aalborg, Denmark. He has worked as Senior Scientist with the Innovation Laboratory of Tata Consultancy Services. His research interests include cross-layer optimization of mobile broadband cellular networks, 5G, Broadband Mobile Communications, 5G Waveform design GFDM FBMC UFMC, heteogeneous networks Femto Cells Device to Device communication, Multi objective optimization for radio access networks, Green radio network design Packet Scheduling and radio resource allocation with link adaptation, MIMO communications. He has delivered several tutorials and seminars on next generation wireless communications. He has guided several PhD students, published several research papers in international journals and conferences. He has co-authored two books titled Adaptive PHY-MAC Design for Broadband Wireless Systems and Single- and Multi-Carrier MIMO Transmission for Broadband Wireless Systems. He has developed teaching resource fading channel and mobile communications freely available as interactive web material for learners of mobile communications. He has taught several subjects such as Modern digital communication techniques, Broadband access systems, mobile communications and fading, teletraffic engineering, introduction to wireless communication and MIMO communications. He is currently serving as associate professor at the G. S. Sanyal School of Telecommunications in the Indian Institute of Technology Kharagpur.

Ramjee Prasad is a Professor of Future Technologies for Business Ecosystem Innovation (FT4BI) in the Department of Business Development and Technology, Aarhus University, Denmark. He is the Founder President of the CTIF Global Capsule (CGC). He is also the Founder Chairman of the Global ICT Standardisation Forum for India, established in 2009. GISFI has the purpose of increasing of the collaboration between European, Indian, Japanese, North-American and other worldwide standardization activities in the area of Information and Communication Technology (ICT) and related application areas.

He has been honored by the University of Rome "Tor Vergata", Italy as a Distinguished Professor of the Department of Clinical Sciences and Translational Medicine on March 15, 2016. He is Honorary Professor of University of Cape Town, South Africa, and University of KwaZulu-Natal, South Africa. He has received Ridderkorset af Dannebrogordenen (Knight of the Dannebrog) in 2010 from the Danish Queen for the internationalization of top-class telecommunication research and education.

He has received several international awards such as: IEEE Communications Society Wireless Communications Technical Committee Recognition Award in 2003 for making contribution in the field of "Personal, Wireless and Mobile Systems and Networks", Telenor's Research Award in 2005 for impressive merits, both academic and organizational within the field of wireless and personal communication, 2014 IEEE AESS Outstanding Organizational Leadership Award for: "Organizational Leadership in developing and globalizing the CTIF (Center for TeleInFrastruktur) Research Network", and so on.

He has been Project Coordinator of several EC projects namely, MAGNET, MAGNET Beyond, eWALL and so on. He has published more than 30 books, 1000 plus journal and conference publications, more than 15 patents, over 100 PhD Graduates and larger number of Masters (over 250).

Several of his students are today worldwide telecommunication leaders themselves.

Under his leadership, magnitudes of close collaborations are being established among premier universities across the globe. The collaborations are regulated by guidelines of the Memorandum of Understanding (MoU) between the collaborating universities.